THE COMPLEX

THE COMPLEX

HOW THE MILITARY
INVADES OUR EVERYDAY LIVES

NICK TURSE

faber and faber

First published in the United States in 2008
by Metropolitan Books, Henry Holt and Company, LLC,
175 Fifth Avenue, New York, New York 10010

First published in the UK in 2008
by Faber and Faber Limited
3 Queen Square London WC1N 3AU

Printed in England by Mackays of Chatham, plc

A CIP record for this book
is available from the British Library

ISBN 978–0–571–22819–5

2 4 6 8 10 9 7 5 3 1

For Tam

CONTENTS

V. THE COMPLEX GOES RECRUITING

VI. THE MAD, MAD WORLD OF THE MILITARY

VII. WEAPONIZING THE FUTURE

VIII. COMPLEX CONCLUSIONS

THE COMPLEX

INTRODUCTION:
A DAY IN THE LIFE

Rick is a midlevel manager in a financial services company in New York City. Each day he commutes from Weehawken, New Jersey, a suburb only a stone's throw from the Big Apple, where he lives with his wife, Donna, and his teenage son, Steven. A late baby boomer, Rick just missed the Vietnam era's antiwar protests, but he's been against the war in Iraq from the beginning. He thinks the Pentagon is out of control and considers the military-industrial complex a danger to the country. If you asked him, it's a subject on which he would rate himself as knowledgeable. He puts effort into educating himself on such matters. He reads liberal Web sites, subscribes to progressive-minded magazines, and is a devotee of Jon Stewart's *Daily Show*.

In fact, he has no idea just how deep the Pentagon rabbit hole goes or how far down it his family already is. Rick believes that, despite its long reach, the military-industrial complex is a discrete entity far removed from his everyday life. Now, if this were 1961, when outgoing President Dwight D. Eisenhower warned the country about the "unwarranted influence" of the "military-industrial complex" and the "large arms industry" already firmly entrenched in the United States, Rick might be right. After all, he doesn't work for one of the Pentagon's corporate partners, like arms maker Lockheed-Martin. He isn't in the Army Reserve. He's never attended a performance of the Marine Corps band (not to mention the army's, navy's, or air force's music groups). But today's geared-up, high-tech Complex is nothing like the olive-drab outfit of Eisenhower's day: It reaches deeper into American lives and the American psyche than Eisenhower could ever have imagined. The truth is that, at every turn, in countless, not-so-visible ways Rick's life is wrapped up with the military.

So wake up with Rick and sample a single autumn morning as the alarm on his Sony (Department of Defense contractor) clock interrupts his final dream of the night. Donna is already up and dressed in fitness apparel by Danskin (a Pentagon supplier that received over $780,000 in DoD dollars in 2004 and another $456,000 in 2005) and Hanes Her Way (made by defense contractor and cake seller the Sara Lee Corporation, which took in over $68 million from the DoD in 2006). Committed to a healthy lifestyle, she's wearing sneakers from (DoD contractor) New Balance and briskly jogging on a treadmill made by (DoD contractor) True Fitness Technology.

Rick drags himself to the bathroom (fixtures by Pentagon contractor Kohler, purchased at defense contractor Home Depot). There, he squeezes the Charmin, brushes with Crest toothpaste, washes his face with Noxzema, then, hopping into the shower, lathers up with Zest and chooses Donna's Herbal Essences over Head & Shoulders—"What the hell," he mutters, "I deserve an organic experience." (The manufacturer of each of these products, Procter & Gamble, is among the top one hundred defense contractors and raked in a cool $362,461,808 from the Pentagon in 2006.)

In go his (DoD supplier) Bausch and Lomb contact lenses and down goes a Zantac for his ulcer (from DoD contractor Glaxo-SmithKline). Heading back to the bedroom, he finds Donna finished with her workout and making the bed—with the TV news on—and lends her a hand. (Their headboard was purchased from Thomasville Furniture, the mattress from Sears, the pillows were made by Harris Pillow Supply, all Pentagon contractors.) They exchange grim glances as, on their Samsung set (another DoD contractor) the *Today Show* chronicles the latest in chaos in Iraq.

"Thank god we never supported this war," Rick says, thinking of the antiwar rally Donna and he attended even before the invasion was launched. (NBC, which produces the *Today Show,* is owned by General Electric, the fourteenth-largest defense contractor in the United States, to the tune of $2.3 *billion* from the DoD in 2006, and has worked on such weapons systems as the UH-60 Blackhawk helicopters and F/A-18 Hornet multimission fighter/attack aircraft, both in use in Iraq.)

Of course, the Pentagon has long poured U.S. tax dollars into private coffers to arm and outfit the military and enable it to function. At the time of Eisenhower's farewell address, *New York Times* reporter Jack Raymond noted that the Pentagon was spending "$23,000,000,000 a year for services and procurement of guns, missiles, airplanes, electronic devices, vehicles, tanks, ammunition, clothing and other military goods." Today, that would equal around $200 billion. In 2007, the Department of Defense's stated budget was $439 billion. Counting the costs of its wars in Iraq and Afghanistan, the number jumps to over $600 billion. Factoring in all the many related activities carried out by other agencies, actual U.S. national security spending is nearly $1 trillion per year.

Back in Eisenhower's day, arms dealers and megacorporations, such as Lockheed and General Motors, held sway over the corporate side of the military-industrial complex. Companies like these still play an extremely powerful role today, but they are dwarfed by the sheer number of contractors that stretch from coast to coast and across the globe. Looking at the situation almost ten years after Eisenhower's farewell speech, Sidney Lens, a journalist and expert on U.S. militarism, noted that there were twenty-two thousand prime contractors doing business with the U.S. Department of Defense in 1970. Today, the number of prime contractors tops forty-seven thousand with subcontractors reaching well over the one hundred thousand mark, making for one massive conglomerate touching nearly ever sector of society, from top computer manufacturer Dell (the fiftieth-largest DoD contractor in 2006) to oil giant ExxonMobil (the thirtieth) to package shipping titan FedEx (the twenty-sixth). In fact, the Pentagon payroll is a veritable who's who of the top companies in the world: IBM; Time-Warner; Ford and General Motors; Microsoft; NBC and its parent company, General Electric; Hilton and Marriott; Columbia TriStar Films and its parent company, Sony; Pfizer; Sara Lee; Procter & Gamble; M&M Mars and Hershey; Nestlé; ESPN and its parent company, Walt Disney; Bank of America; Johnson & Johnson; among many other big-name firms. But the difference between now and then isn't only in scale. As this list suggests, Pentagon spending is reaching into areas of American life previously neglected: entertainment, popular consumer brands, sports. This penetration translates into a remarkable variety of forms of interaction with the public.

Rick and Donna's home is full of the fruits of this incursion. As they putter around in their kitchen, getting ready for the day ahead, they move from the wall cabinets (purchased at DoD contractor Lowe's Home Center) to the refrigerator (from defense contractor Maytag), choosing their breakfast from a cavalcade of products made by Pentagon contractors. These companies that, quite literally, feed the Pentagon's war machine, are the same firms that fill the shelves of America's kitchens.

Like everything else about the military, food has undergone radical changes in recent years. The navy deep-sixed its dairy back in the late 1990s. And huge piles of potatoes peeled by privates on "KP" duty, a joke of old war films, largely went out when the "kitchen patrol" was privatized and became big business for firms like the French food services company Sodexho (which feeds the Marine Corps and received more than $154 million from the DoD in 2006), Agility Logistics, a Kuwait-based company formerly named Public Warehousing (that reaped $1.8 billion from the Pentagon in 2006), which is the U.S. Army's principal food supplier for the Iraq war zone, and KBR (formerly Kellogg Brown and Root), whose employees and subcontractors do everything from cooking the meals to washing the dishes for the U.S. military in Iraq. Today, just about every supermarket staple has ties to the Pentagon.

THE PENTAGON IN YOUR PANTRY

Food	DoD Contractor	Food	DoD Contractor
After Eight mints	Nestlé	Coffee Mate	Nestlé
Ajax Scourer	Colgate-Palmolive	College Inn Chicken Broth	Del Monte Foods
Aquafina purified drinking water	PepsiCo	Contadina tomato paste	Del Monte Foods
Athenos Hummus	Kraft/Altria	Country Crock margarine	Unilever N.V.
Aunt Jemima syrup	PepsiCo	Cream of Wheat	Kraft/Altria
Bagel Bites	H.J. Heinz	Cream-O-Land Milk	Cream-O-Land Dairy
Baked Doritos tortilla chips	PepsiCo	Crystal Light	Kraft/Altria
Balance Bars	Kraft/Altria	Dasani purified, noncarbonated water	Coca-Cola
Ball Park franks	Sara Lee		
Bertolli olive oil	Unilever N.V.	Dawn dishwashing liquid	Procter & Gamble
Birds Eye Asian Vegetables in Sesame Ginger Sauce	Birds Eye	Del Monte diced tomatoes	Del Monte Foods
Birds Eye Voila! Chicken Fajita	Birds Eye	Dixie Crystals sugar	Imperial Sugar
Boboli Pizza Crust	George Weston Bakeries	Dole Canned Pineapples	Dole Food
Bounty paper towels	Procter & Gamble	Earth Grains Frozen Garlic Bread	Sara Lee
Breakstone's Cottage Cheese	Kraft/Altria	Egg Beaters pasteurized egg product	ConAgra Foods
Buffalo Style Popcorn Chicken Bites	Tyson Foods	Eggo Blueberry Pancakes	Kellogg's
Bumble Bee Tuna	Bumble Bee Seafoods	Eggo waffles	Kellogg's
Butterball Turkey	ConAgra Foods	Entenmann's Crumb Donuts	George Weston Bakeries
C&W Cut Italian Green Beans	Birds Eye	Fig Newtons	Nabisco/Kraft/Altria
Cheerios	General Mills		
Chi-Chi's Salsa	Hormel		
Chunky Vegetable and Pasta Soup	Campbell Soup		

Food	DoD Contractor	Food	DoD Contractor
Fruit$_2$0 fruit-flavored water	Kraft/Altria	Mr. Clean multi-purpose cleaner	Procter & Gamble
Grape-Nuts	Kraft/Altria	Mr. Yoshida's Hawaiian Sweet & Sour Sauce	H.J. Heinz
Green Giant Asparagus Spears	General Mills	Mueller's pasta	American Italian Pasta
Grey Poupon Dijon mustard	Kraft Bakeries	Muir Glen Classic Minestrone soup	General Mills
Heinz Vinegar	H.J. Heinz	Natural Ruffles potato chips	PepsiCo
Herb-Ox bouillon	Hormel Foods	Nature Valley granola bars	General Mills
Hershey's Chocolate syrup	Hershey Foods	Naturewell antibiotic-free, corn-fed beef	National Beef Packing
Hillshire Farm Smoked Turkey Breast	Sara Lee	Near East Couscous	PepsiCo
Horizon Organic Milk	Dean Foods	Nestea Sweetened Lemon Iced Tea	Nestlé
Ice Breakers chewing gum	Hershey Foods	Nissin Cup Noodles	Nissin Foods
Jack Daniel's Grilling Sauce	H.J. Heinz	Ocean Spray cranberry juice	Ocean Spray Cranberries, Inc.
Jimmy Dean Low Sodium Premium Bacon	Sara Lee	Old El Paso salsa	General Mills
Kikkoman teriyaki marinade	Kikkoman International	Orbit Gum	Wm. Wrigley Jr.
Kix cereal	General Mills	Ore-Ida french fries	H.J. Heinz
La Choy soy sauce	ConAgra Foods	Oreo cookies	Nabisco/Kraft/Altria
Lipton Iced Tea	Unilever N.V.	Organic Baby Spinach	Dole Food
Marcal paper napkins	Marcal Paper Mills	Palmolive dish soap	Colgate-Palmolive
McCormick Pure Vanilla Extract	McCormick		
Minute Maid Light Lemonade	Coca-Cola		

Food	DoD Contractor	Food	DoD Contractor
Pepperidge Farm Seven Grain bread	Campbell Soup	Softsoap Naturals liquid hand soap	Colgate-Palmolive
		Special K cereal	Kellogg's
Poland Spring water	Nestlé	Spoon Size Shredded Wheat	Kraft
Powerbars	Nestlé	Stouffer's Grilled Vegetable French Bread Pizza	Nestlé
Prilosec OTC heartburn medication	Procter & Gamble		
Progresso New England Clam Chowder	General Mills	Swanson Certified Organic Vegetable broth	Campbell Soup
		Thomas' English Muffins	George Weston Bakeries
Puffs facial tissue	Procter & Gamble		
Quaker Oatmeal	PepsiCo	Toblerone chocolate	Kraft/Altria
Rice Krispies	Kellogg's	Tropicana orange juice	PepsiCo
Ronzoni pasta	New World Pasta		
S&W black beans	Del Monte Foods	Tyson Chicken Broth	Tyson Foods
San Giorgio elbow macaroni	New World Pasta	Tyson Chicken Fajitas Meal Kit	Tyson Foods
Silk Soymilk	Dean Foods		
Skippy Natural peanut butter	Unilever N.V.	Uncle Ben's rice	Mars, Inc.
		V8 vegetable juice	Campbell Soup
Slim•Fast Meal On-The-Go	Unilever N.V.		
Smucker's Natural Peanut Butter	J.M. Smucker	Wheat Thins	Nabisco/Kraft/Altria
		Wheatables crackers	Keebler
Smucker's strawberry preserves	J.M. Smucker	Wish-Bone Italian salad	Unilever N.V.
Snackwell's Fat Free Devil's Food Cookie Cakes	Kraft	Wolfgang Puck's gourmet pizza	ConAgra Foods

The same holds for many household appliances. In Rick and Donna's dining room, a small Roomba robotic vacuum cleaner buzzes around the floor. Rick thought it would be cute to have the little mechanical device trolling around the house making their hectic lives just a tad easier. Little did he know that Roomba's manufacturer, iRobot, takes in U.S. tax dollars ($51 million of them from the DoD in 2006, more than a quarter of the company's revenue) and turns them into PackBots, tactical robots used by U.S. troops occupying Iraq and Afghanistan, and, beginning in 2008, Warrior X700s—250-pound semiautonomous robots armed with heavy weapons such as machine guns. In addition to selling millions of Roombas to civilian consumers, the company uses government tax dollars to make money on the civilian side of its business. According to the company's December 2006 annual report (which listed as its "Research Support Agencies" the Defense Advanced Research Projects Agency [DARPA], the U.S. Space and Naval Warfare Systems Command, the U.S. Army Tank-Automotive and Armaments Command, and the U.S. Army Armament Research, Development and Engineering Center), government funding "allows iRobot to accelerate the development of multiple technologies." Yet iRobot retains "ownership of patents and know-how and [is] generally free to develop other commercial products, including consumer and industrial products, utilizing the technologies developed during these projects." It's a very sweet deal. And iRobot is hardly alone.

Sitting on the dining room table is Rick's HP (Hewlett-Packard) notebook computer. HP is another company that has grown its civilian know-how with generous military contracts, like the multiyear, multimillion-dollar deal it signed, in 2005, with DARPA to "develop technologies to improve the performance of mission-critical computer networks used during combat and other vital operations." A spokesman for the company noted, "Our work for DARPA is aimed at significantly improving the performance of the Internet. . . . If we can successfully create new approaches to the way Internet traffic is detected and routed, we may start seeing the Internet used as the de facto communications and information network in areas where it previously would've been thought too risky." Success would certainly translate into more lucrative civilian work, as well.

Meanwhile, Rick and Donna's son, Steven, is still upstairs, having a hard time tearing himself away from his computer game. His room is a veritable showcase of the new entertainment-sports-high tech-pop culture dimension of the twenty-first-century Complex: there are NASCAR posters (in 2005, more than $38 million in taxpayer money was spent on U.S. armed forces' racecars); National Football League (NFL) jerseys and baseball caps (the NFL has partnered with the Pentagon to create military profiles aired during TV broadcasts of regular and postseason games, while individual NFL teams have hosted "military appreciation" events); X-Men comic books (the Pentagon teamed up with Marvel Comics to produce limited-edition "military-exclusive" comic books, with pro-Pentagon themes, that are now sought after by civilian collectors), and a wastebasket filled with empty Mountain Dew bottles (the air force was one of the sponsors of the Dew Action Sports Tour, a traveling show featuring skateboarding, BMX, and freestyle motocross contests).

During Ike's time, when firms like Ford and AT&T were the big military suppliers—the payroll showed an utter lack of cool companies. Now, the Pentagon is reaching into virgin territory in new ways with new partners. Today, hip firms like Apple, Google, and Starbucks are also on DoD contractors' lists. And while Ike's complex was typified by brass bands and patriotic parades, today's variant is a flashy digitized world of video games, extreme sports, and everything cool that appeals to potential young recruits.

Steven finally shuts down Tropico: Paradise Island—a nation-building simulation video game where the player, as "El Presidente," attempts to lure tourists to his/her fun-in-the-sun resort. Neither father nor son is remotely aware that the software maker, Breakaway Games, does taxpayer-funded work for such military clients as DARPA, the Joint Forces Command, the Office of the Secretary of Defense, and the United States Air Force—as well as having developed 24 Blue, a simulator used to improve aircraft carrier–based operations. They are blissfully unaware of even the existence of Breakaway's Pentagon-funded video game that could conceivably lead to more effective bombing of targets abroad. Steven grabs his iPod MP3 player (from DoD contractor Apple Computer) and heads downstairs to leave with his father. Heading for the door, Rick goes to his bookshelf and scans a selection of progressive texts whose publishers just happen to be DoD contractors, including a reissue of Rachel Carson's *Silent Spring* (Houghton Mifflin), *Bushwhacked: Life in George W. Bush's America* by Lou Dubose and Molly Ivins (Random House), and Jon Stewart's *America (The Book)* (Warner Books), before choosing the Hugo Chavez–approved *Hegemony or Survival* by Noam Chomsky (*ahem*, Metropolitan Books from Macmillan publishers). As the last one out, Donna sets the ADT alarm system (ADT took in more than $16 million from the Pentagon in 2006, while its parent company, Tyco International, cleaned up to the tune of over $187 million).

Rick and Steven hop into the Saturn parked in the driveway. Rick is proud of his car choice—after all Saturn has such a people-friendly (even anti–Detroit establishment) vibe. Admittedly, he is aware that General Motors owns not only the Saturn but the Hummer brand—the civilian version of the U.S. military's Humvee—but he believes that in this world, you can't be squeaky-clean perfect. But Hummer isn't the half of it.

How could Rick have known that in 1999, GM formally entered the army's COMBATT (COMmercially BAsed Tactical Truck) vehicle development program? Or that GM actually had its own military division, General Motors Defense, when his Saturn was made? Nor could Rick have known that GM Defense formed a joint venture with defense giant General Dynamics to create the GM-GDLS Defense Group (which was awarded in excess of $1.5 *billion* in DoD contract dollars in 2005), or that GM took in $87 million from the Pentagon in 2006. Or that, in 2007, GM entered into a fifty-year lease agreement to build a $100 million test track on the U.S. Army's Yuma Proving Grounds. Or that the maker of his Saturn's tires, Goodyear, was America's sixty-ninth-largest defense contractor in 2004, with DoD contracts worth nearly $357 million.

Rick might be an aging baby boomer, but he still tries to look cool (to Steven's embarrassment). As he pulls the Saturn out of the driveway, he dons a pair of Oakley sunglasses. Oakley supplies goggles and boots to U.S. troops. And while the military purchased goggles from firms such as the American Optical Company during the 1940s, it's unlikely that anyone ever called that company's designs "badass," as *Powder,* a skiing magazine that runs army recruitment ads on its Web site, called one of Oakley's products. Driving along, Rick glances over at his son. "Are those the Wolverine boots we just got you?" "Yeah, Dad," answers Steven, looking down at his now-ratty footwear. Rick's already thinking about the next pair he'll need to buy his son, not about the five-year, multimillion-dollar contract the company signed in 2003 to supply the army with an upgraded infantry combat boot, or the other deals, worth tens of millions of dollars, that Wolverine signed with the Pentagon in 2004, 2006, and 2007.

As they drive to his school, Steven perks up. "That's it, Dad!" he says, pointing at a Ford Escape that just pulled into the high school parking lot. "Whaddaya say, Dad? Next year, when I get my license?" Rick remembers hearing on the radio that Ford makes an Escape hybrid-electric vehicle. "You know what, son? I think maybe we just might look into it." He experiences a little burst of satisfaction. Not only can he feel like a *good* dad but as a bonus he can even help the environment. (Ford Motor Company and its subsidiaries have, of course, garnered rafts of defense contracts and aided the army and navy in various projects.)

Overjoyed, Steven shoots his father a big smile as he opens the car door, "Alright! Well, I'll see you tonight, Dad." "Do you have your cell phone?" Rick asks. Steven whips a Motorola from his pocket. (Motorola made almost $308 million from the Department of Defense in 2004, while the phone's service provider, Verizon, took home more than $128 million in DoD contracts, and $50 million more from the Department of Homeland Security, in 2006.)

With Steven at school, Rick heads for work. He gives the local Exxon station (ExxonMobil took in more than $1.17 *billion* in

DoD dollars in 2006) a pass and instead pulls into Shell, which likes to portray itself as a kinder, greener oil giant. As he signs the receipt of his Bank of America credit card (a firm which issues special credit cards to Pentagon employees to streamline the process of buying supplies for the DoD), Rick has no way of knowing that Shell's parent company, N.V. Koninklijke Nederlansche, was the thirty-first-largest defense contractor in 2006, reaping over $1.15 *billion* dollars in DoD contracts.

Rick's route to work doesn't pass an army barracks, an air force base, Halliburton's offices, or a single factory of a defense giant like Boeing or General Dynamics. Nevertheless, he manages to zip by a veritable who's who of defense contractors or their subsidiaries even if they look like the staple businesses of Anytown, USA. They include:

an AMF bowling alley
a Wal-Mart
an OfficeMax office supply store
an Ace Hardware store
a CompUSA computer store
a Hilton hotel
an Avis Rent-a-Car
a Staples office supply store
a Home Depot
and
a BP gas station

Entering the Holland Tunnel on his way to Manhattan, Rick realizes that, with Steven driving next year, he can start taking mass transit to work. The PATH train into the city—recently restored under the watchful eye of Bechtel, the fifteenth-largest defense contractor of 2004 and the recipient of more than $1.7 *billion* in DoD contracts that year—will, he believes, lessen his "footprint" on the planet.

Keep in mind, Rick is now only a couple of hours into his long day. In fact, no part of the hours to come will be lacking in products produced by Pentagon contractors—from the framed photographs of Donna and Steven on his desk (taken by an Olympus camera and printed on Kodak paper) to the beer he drinks with lunch (Budweiser) to most of the products around his office, including: 3M Post-It notes, Microsoft Windows software, Lexmark printers, Canon photocopiers, AT&T telephones, Maxwell House Coffee from Altria, Kidde fire extinguishers, Xerox fax machines, IBM servers, paper from International Paper, Duracell batteries from Procter & Gamble, an LG Electronics refrigerator, and paper towels by Marcal Paper Mills.

Later in the day, Rick drops in to see the head of the company's IT department and, gazing around the room, notices the name on some software: Oracle, "the world's largest enterprise software company." Even if he knew that Oracle received over $93 million from the DoD in 2005, what would remain invisible is the way many corporations that supply the military additionally belong to a tight-knit, mutually reinforcing clan of contractors. For example, not only do "7 of the top 10 aerospace and defense companies run Oracle applications," including top arms dealers Boeing ($20.3 *billion* in sales to the DoD in 2006) and General Dynamics, but so do other tech-industry heavy hitters such as IBM (which sold the navy the "fastest supercomputer in the US military" in 2004). Additionally, the company has partnered with Microsoft (which, in 2006, received over $41 million from the DoD) and Sun Microsystems, whose High Productivity Computing project is partially funded by a $50 million award from DARPA. Additionally, Oracle's client General Dynamics also counts among its "strategic partners" Alcatel Worldwide, Qwest, and Cisco Systems—all of them inextricably bound up in "the Complex." In 2004, Qwest—"a leading provider of voice, video and data services" to more than 25 million customers—inked a five-year deal to be "the exclusive provider of data and voice communications to the White Sands Missile Range." Dizzy yet? It gets worse. Cisco Systems, a "worldwide leader in networking for the Internet," has "strategic alliances" with DoD contractors, such as Microsoft, IBM, and Intel, and signed a deal in 2004 with that year's number two Pentagon contractor Boeing (which makes Harrier II tactical strike aircraft for the marines) to enhance the defense giant's voice and data networks. But this is just business as usual in the incestuous tangle of Pentagon contractors.

After a long day at the office, Rick heads home ready to catch NBC *Nightly News* and later *Law and Order* (both products of defense giant General Electric). Or maybe he'll tune in to CBS and watch *NCIS* (Naval Criminal Investigative Service), the action-drama starring Mark Harmon that has received assistance from the U.S. military. Or, since the whole family will be home, they might even play Scrabble. While it's decidedly low tech and far from *cool,* even Scrabble has ties to the Pentagon. The game's maker, Hasbro, has had a long relationship with the military and whenever they need specs for an action-figure's uniform or some other insider info they just pick up the phone or send an e-mail—as they did to the Army Soldier Systems Center, two days after the Iraq War began, in order to get the latest intel on military gear.

Rick is, of course, a fiction, but the rest of us aren't—and neither is the existence of the real Matrix.

In 1957, the prescient scholar C. Wright Mills pointed to World War II as the moment when the "merger of the corporate economy and the military bureaucracy came into its present-day significance." Mills's identification of an "economic-military alliance"— reaching into the halls of politics, science and technology, and higher education—presaged Eisenhower's 1961 address warning about the growth of a full-blown American "military-industrial complex."

In that capstone speech to the nation, the departing commander in chief explained to the American public that the United States had created "a permanent armaments industry of vast proportions" and that the perils of this new "conjunction of an immense military establishment and a large arms industry" were dire. Warning against "the acquisition of unwarranted influence, whether sought or unsought, by the military industrial complex," he insisted that "only an alert and knowledgeable citizenry can compel the proper meshing of the huge industrial and military machinery of defense with our peaceful methods and goals, so that security and liberty may prosper together."

By the time of Eisenhower's farewell address, the military-industrial complex was already well entrenched in American life and the public was not up to the task of checking, let alone reversing, its power—especially in the Cold War world. Today, it's nearly impossible to imagine the public even considering the task, no less imagining what it would require. After all, even Ike never imagined the emergence of a Complex of such epic proportions that it would someday almost entirely envelop American culture.

However, the military-industrial complex was never truly confined to the "armaments industry," as Mills and others made perfectly clear. As early as the 1960s, some scholars began to write about a "military-industrial-academic complex" or a "Golden Triangle" of "military agencies, the high technology industry, and research universities." Others focused on the "Iron Triangle" of military contractors, the Pentagon, and Congress. Still others have proposed such self-explanatory formulations as the "military-industrial-media-entertainment network," "military-industrial-entertainment complex," "military-industrial-think tank complex," or even the "metropolitan-military complex." In recent years, commentators have added whole new associated subcomplexes like the "security-industrial complex," the "homeland security complex," the "cybersecurity industrial complex," the "surveillance-industrial complex," as well as allied entities like the prison-industrial complex, the disaster-capitalism complex, and various other adjuncts to and derivations from Eisenhower's now-famous formulation. None of these, however, completely describes, let alone captures, the full breadth of the Complex as it exists today.

Identifying the precise parameters of the Complex has proven as difficult as precisely naming it. In *Imperial Delusions,* Carl Boggs, an expert on U.S. militarism, locates America's "new militarism" at the "intersection of globalization and the entrenchment of U.S. economic, political and military power—a matrix well beyond the

question of who specifically is involved in elite decision making."
Although the word slips by almost unnoticed, Boggs's reference to
a "matrix" is right on the money.

In the 1999 sci-fi movie classic of the same name, the Matrix is
an artificial reality (resembling the Western world at the dawn of
the twenty-first century) created by sentient machines. Humans,
who are grown as energy sources and wired in to the Matrix using
cybernetic implants, are kept in a comalike state—ignorant of the
very existence of the artificial reality that they "live" in. In explaining the situation to Neo, the movie's protagonist, Morpheus, a
leader of a group of *unplugged* free humans who wage a guerrilla
struggle against the machines, reveals:

> The Matrix is everywhere. It is all around us. Even now, in this
> very room. You can see it when you look out your window or
> when you turn on your television. You can feel it when you go to
> work, when you go to church, when you pay your taxes. It is the
> world that has been pulled over your eyes to blind you from the
> truth.

At one point in his farewell speech, Eisenhower even presaged
this point, suggesting, "The total influence—economic, political,
even spiritual—[of the conjunction of the military establishment
and the large arms industry] is felt in every city, every State house,
every office of the Federal government." But only Hollywood has
managed to capture the essence of today's omnipresent, all-
encompassing, cleverly hidden system of systems that invades all
our lives; this new military-industrial-technological-entertainment-
academic-scientific-media-intelligence-homeland security-surveil-
lance-national security-corporate complex* that has truly taken
hold of America.

The Complex is connected to everything you would expect,
from the top arms manufacturers to big oil corporations—as well

---

*Obviously, this name is too unwieldy for general use. Therefore, this vast system
of systems will throughout this book be referred to as "the Complex," the
"military-corporate complex," or "the real Matrix."

as numerous government agencies connected to the U.S. Department of Defense and allied entities such as the Central Intelligence Agency and the Department of Homeland Security. But it is also connected to the entertainment industry and the world's largest media conglomerates. It is in league with the nation's largest food suppliers and beverage companies. It supports the most prestigious universities in America and is tied to the leading automakers.

This symbiotic relationship is not limited to megacorporations. Thousands and thousands of small-town niche contractors have their hands out as well. In 2004, such lesser lights included: Kenny's Liquor of Parkston, South Dakota, and Chic-A-D's Cajun Chicken & Catfish Restaurant in Winnsboro, Louisiana. In 2005, some of the less-than-likely small fry on the growing list were the Maxfield Candy Company of Salt Lake City, Utah; the High Sierra Toilet Company of New Braunfels, Texas; the U.S. Toy Company of Grandview, Missouri; Butt Construction Company of Springfield, Ohio; Corky's Bar-B-Que of Memphis, Tennessee; and Skip's Sports Equipment of Virginia Beach. In 2006, the Pentagon payroll also included the New Pig Corporation of North Little Rock, Arkansas; the Spangler Candy Company of Bryan, Ohio; Safari Country Paintball of Lewton, Oklahoma; the Colorado Boxed Beef Company of Auburndale, Florida; Gilbert's Egg Inc. of Forest, Mississippi; and the One Way Christian Book Store of Hattiesburg, Mississippi.

Most striking in this new age of corporate-military-entertainment meldings are those by-products of the Complex's effort to project a cool, hip image, including military-crafted simulators that have become commercial video games; NASCAR events that feature race cars sponsored by branches of the armed forces; slick recruiting campaigns that use the hottest social networking technology to capture the attention of teens; and involvement with civilian outfits popular with the young, like Starbucks, Oakley, Disney, and Coca-Cola. Just like the fictional Matrix, the Complex is nearly everywhere and involved in almost everything, and very few people aren't plugged into it in some way, shape, or form. Above all, as in the movie, most people are hardly aware that this "real Matrix" even exists.

Today's military-corporate complex is nothing if not sophisti-cated. It uses all the tools of the modern corporation: publicity departments, slick advertising campaigns, and public relations efforts to build up the armed forces, which are, of course, its raison d'être. With an all-volunteer military embroiled in two disastrous wars in Afghanistan and Iraq, the armed forces have had to ramp up advertising, marketing, and product-placement efforts to attract ever more reluctant new recruits. At the same time, those not won over to actual military service are called upon—whatever their feelings about specific wars—to serve in other ways, through their tax dollars, most obviously, but also by working for corporations that fuel the military and are fueled by it, as well, of course, as by ensuring the economic well-being of these firms through their purchases. The high level of military-civilian interpenetration in a heavily consumer-driven society means that almost every Ameri-can (aside, perhaps, from a few determined anarcho-primitivists) is, at least passively, supporting the Complex every time he or she shops for groceries, sends a package, drives a car, or watches TV— let alone eats barbeque in Memphis or buys Christian books in Hattiesburg. And what choice do you have? What other brand of computer would you buy? Or cereal? Or boots?

In *The Matrix,* Neo looks at the Matrix's signature code cascad-ing down a computer monitor and asks Cypher, a human who eventually betrays the guerrilla band, if it always must be viewed encoded. Cypher replies, "Well you have to . . . there's way too much information to decode the Matrix. You get used to it. I don't even see the code." This book is intended to decode at least some parts of the real Matrix and make them visible. Think of this book as the equivalent of *The Matrix*'s "red pill"—a primer to introduce you to the new world of the military-corporate complex . . . your world.

PART I

KICKING IT
OLD SCHOOL

Before we embark on an exploration of the Complex's assortment of unlikely defense contractors and strange sectors of interest, it's important to have a picture of the workings of the old, Eisenhower-style military-industrial complex, which has only strengthened since Ike's famous address. So let's take a moment to explore the interplay of arms manufacturers, the Pentagon, and the U.S. Congress—the "iron triangle," as scholars have long called it. From there, we'll consider one of the Pentagon's traditional stomping grounds—its ol' alma mater—the campuses of America's colleges and universities. Then it's on to the "military-petroleum complex," which demonstrates that while antiwar activists may chant "no blood for oil," the Pentagon's mantra might well be "more oil for blood." Finally, we'll take a spin through the military's many bases and even all-military resorts to get a sense of just how much of this planet the Pentagon has really put its footprint on.

1

THE IRON TRIANGLE

Before the Complex of today came into existence there were the immensely powerful arms manufacturers of Eisenhower's military-industrial complex. They haven't exactly gone away. During Eisenhower's last term in office, from 1957 to 1961, the top five military contractors were General Dynamics, Boeing, Lockheed, General Electric, and North American Aviation. These days, General Electric has slipped slightly (it ranked fourteenth among contractors in 2006), and what's left of North American Aviation, sold and resold over the years, now exists as part of United Technologies (the ninth-largest contractor in 2006). But even with massive consolidation—a 2003 Pentagon report found that the fifty largest defense contractors of the early 1980s have become today's top five contractors—the leading players remain largely the same.

In 2002, the massive defense contractors Lockheed Martin, Boeing, and Northrop Grumman ranked one, two, and three among the Pentagon's defense contractors, taking in $17 billion, $16.6 billion, and $8.7 billion, respectively. Lockheed, Boeing, and Northrop Grumman did it again in 2003 ($21.9, $17.3, and $11.1 billion); 2004 ($20.7, $17.1, and $11.9 billion); 2005 ($19.4, $18.3, and $13.5 billion), and 2006 ($26.6, $20.3, and $16.6 billion). General Dynamics ranked fourth in four of the five years—and never lower than fifth.

For decades, these military-industrial powerhouses have produced some of the most sophisticated and deadly weaponry in the U.S. arsenal—from the U-2 spy-plane and Javelin missile (Lockheed) to the B-52 Stratofortress heavy bombers that pounded Indochina (Boeing) to the famed B-2 stealth bomber (Northrop Grumman). Over the years, they have also grown in power and influence, often bending Congress and presidential administrations to their lobbying will and, in some cases, dwarfing the financial clout of various arms of the government itself. For example, from 2000 to 2006, Lockheed Martin received $135.4 billion from the Pentagon—in addition to contracts with the Departments of Homeland Security, Justice, and Commerce, the Federal Aviation Administration, and the National Aeronautics and Space Administration (NASA), among others. In 2005, the Bethesda, Maryland–based company's $25 billion in federal contracts exceeded the total combined budgets of the Departments of Commerce and the Interior, the Small Business Administration, and the U.S. Congress. Peter Singer, an expert on military privatization, uses such examples to suggest that some defense contractors have evolved beyond any conceivable definition of a private business concern. "They're not really companies, they're quasi agencies," he told the *New York Times*. That's certainly the case with Lockheed, whose annual sales to U.S. government agencies clocked in at 78 percent of the company's total business in 2003, 80 percent in 2004, and 85 percent in 2005. In 2006, there was a slight dip to 84 percent, but that didn't count 13 percent of sales to foreign governments, "including foreign military sales funded, in whole or in part, by the U.S. Government."

But it isn't only percentages of sales that link companies like Lockheed so closely to the Pentagon; it's also their long history of aiding the military in various wars, interventions, incursions, and invasions. The corporate forebears of the firm—which was formed in 1995 when two defense industry stalwarts, Lockheed Corporation and Martin Marietta Corporation, merged—have been working for and with the army and navy since the early years of the twentieth century. And the Complex's "revolving door," through which arms manufacturers' employees and DoD officials routinely passed back and forth between the public and private realm, has continued to turn, creating a familiarity that has cemented the relationship, while offering generous rewards to all involved. (In 1960, at the end of the Eisenhower era, "726 former top ranking military officers were employed by the country's 100 leading defense contractors.")

In 2001, to take but one example, Edward C. "Pete" Aldridge Jr. left the defense giant Aerospace Corporation (which took in more than $442 million from the DoD that year) to become the undersecretary of defense (acquisition, technology, and logistics) in Donald Rumsfeld's Pentagon. Only months into his tenure, Aldridge chose Lockheed Martin to build the F-35 Joint Strike Fighter. In an interview with Ron Insana on CNBC, he announced, "It is the largest acquisition program in the history of the Department of Defense. We're expecting it to be in excess of $200 billion over the period of the program." Only minutes later, Aldridge spoke to CNN's Lou Dobbs, who asked, "How soon will the money start moving from the federal government to Lockheed Martin?" Aldridge replied, "I think Lockheed Martin would like to see some right away . . . as soon as the work starts, they can start billing the government for the work that they're producing. And so I would expect it's a matter of a few months away. But very soon." Lockheed would, indeed, soon be getting paid. By 2006, the estimated cost of the project had already jumped from $200 billion to $276 billion.

In early 2003, despite having previously criticized the program as too expensive, Aldridge approved a $3-billion contract for Lockheed's controversial F-22 Raptor fighter jet. Then, in March, he announced his retirement, saying, "Now it is time, for personal reasons, to move on to a more relaxed period of my career. I will continue to support the national security interests of this country, albeit in a less direct way." Within months, Aldridge had been elected to Lockheed Martin's board of directors, complete with six-figure compensation and company stock, while a former Lockheed man and air force veteran, Michael W. Wynne, took over his post (and went on to become secretary of the air force in 2005).

By 2007, Aldridge was still on the board and held 5,041 units of the corporation's common stock. The deals he made with Lockheed seem to have paid off handsomely. When Aldridge joined the Pentagon in May 2001, Lockheed's stock was trading at about $36.62 per share; in 2003, about the time he was elected to Lockheed's board, it was at $47.52, and by late October 2007 it was trading at $108.61.

Aldridge's move to Lockheed raised some eyebrows but it's not clear why, given his long history of using the revolving door. In the 1960s, he had held various posts with Douglas Aircraft Company's Missile and Space Division. In 1967, the year Douglas merged with McDonnell Aircraft (today, McDonnell Douglas is a wholly owned subsidiary of Boeing), he joined the staff of the assistant secretary of defense for systems analysis and stayed in the post until 1972. He then became a senior manager with LTV Aerospace Corporation before heading back to the government, from 1974 to 1976, to serve as deputy assistant secretary of defense for strategic programs. Dizzyingly enough, in 1977, he was back in the private sector as vice president of the National Policy and Strategic Systems Group for the System Planning Corporation, a major defense contractor. In 1981, Aldridge spun back the other way to become undersecretary of the air force, a position he held until 1986. (From 1981 to 1988, he also served as the director of the U.S. National Reconnaissance Office.) In 1986, he became the secretary of the air force and held the post until December 1988, when he became the president of defense giant McDonnell Douglas's Electronic Systems Company—the post he held until leaving to join the Aerospace Corporation in 1992.

Aldridge's revolving-door escapades were far from unique. In 2004, the Project on Government Oversight, an independent nonprofit group that investigates and exposes corruption, reported that, between January 1997 and May 2004, at least 224 senior government officials had taken top positions with the twenty largest military contractors. Lockheed headed the list with thirty-five lobbyists, sixteen executives, and six directors or board members—a total of fifty-seven former senior government officials who had crossed over to the other side. In 2007, for example, in addition to Aldridge, the Lockheed board boasted General Joseph W. Ralston, who had served as vice chairman of the Joint Chiefs of Staff from 1996 to 2000; James O. Ellis Jr., a navy veteran who had retired from his post as an admiral and commander, U.S. Strategic Command, in July 2004; Admiral James Loy, a former commandant of the U.S. Coast Guard who retired in 2005, after serving as the first deputy secretary of homeland security; Eugene F. Murphy, a Marine Corps veteran and former attorney with the Central Intelligence Agency; and Robert J. Stevens, the chairman, president, and chief executive officer of Lockheed Martin, who had served in the U.S. Marine Corps and is a graduate of the Department of Defense Systems Management College Program Management course.

Under President George W. Bush and his first secretary of defense Donald Rumsfeld, the Pentagon was especially well connected to Lockheed. As the author and playwright Richard Cummings noted in *Playboy*, Powell A. Moore, the assistant secretary of defense for legislative affairs, had, from 1983 to 1998, been a consultant and vice president for legislative affairs for Lockheed. Secretary of the navy and later deputy secretary of defense Gordon England had also worked for Lockheed, as had Peter B. Teets, who became the undersecretary of the air force and director of the National Reconnaissance Office, while Albert Smith, Lockheed's executive vice president for integrated systems and solutions, was appointed to the Defense Science Board. And the connections didn't end with the Pentagon. Joe Allbaugh, who moved from being the Bush-Cheney ticket's national campaign manager to the head of the Federal Emergency Management Agency (and brought his college buddy Michael "Brownie, you're doing a heck of a job" Brown into the agency), was a Lockheed lobbyist. Transportation Secretary Norman Y. Mineta, the lone Democrat in Bush's cabinet, had been a Lockheed vice president; Vice President Dick Cheney's son-in-law, Philip J. Perry, was a registered Lockheed lobbyist; and Cheney's wife, Lynne, had served, until 2001, on Lockheed Martin's board of directors.

Of course, any discussion of the classic military-industrial complex would be incomplete without the third side of the "iron triangle": the U.S. Congress. Without the Senate, revolving-door warriors like Aldridge would never get approved for their Pentagon slots, nor would their pet defense programs from past or future defense contractor employers be funded. It's America's legislative representatives who pump up the pork in Washington in order to bring home the bacon for their districts (and themselves). Take Aldridge's baby, the F-22—a fighter designed to counter advanced Soviet aircraft that were never built. Beset by decades of huge cost overruns and embarrassing episodes (such as when a pilot had to be removed from the plane's defective cockpit using a chain saw) as well as numerous delays and setbacks, the fighter-without-a-foe seemed to be a prime candidate for the chopping block. With real-world costs of each of the 183 planes in the program topping out at $350 million and even powerful Republicans on the Senate Armed Services committee, including Arizona senator John McCain and the committee's chairman, Virginia senator John W. Warner, lined up against the Lockheed lobby, it looked like the F-22 might be done for in 2006.

The "powerful F-22 lobby, a combination of the air force, Lockheed Martin, which makes the fighter jet, and their allies in Congress" however were too strong to defeat. As a result, McCain and Warner then attempted to beat back the drive for a multiyear contract for the F-22, which would allow budget cutters a shot at scrapping the program the next year. Prior to the vote, however, Lockheed engaged in a full-scale lobbying effort capped off by an e-mail campaign asking senators to vote "yes" on the proposed Chambliss Amendment, which called for a multiyear deal. In short order, the foes of the F-22 went down in flames. Danielle Brian, the executive director of the Project on Government Oversight, explained, "The F-22 lobby is an extraordinary juggernaut and they fought to the death on this one." She observed, "It is astonishing in that the lobby can take on the most powerful in Washington, including the president, and win."

The Chambliss Amendment was named for Saxby Chambliss, a Republican senator from Georgia whose district just happened to include an F-22 assembly plant. And Chambliss wasn't alone in being locked in to Lockheed. The e-mail campaign reminded many senators where their bread was buttered. In fact, spreading the wealth is one prime way Lockheed does business. In addition to its Marietta, Georgia, plant, Lockheed produces the F-22 Raptor at facilities in Palmdale, California; Meridian, Mississippi; Fort Worth, Texas; and even at a Boeing plant in Seattle, Washington. For added insurance, Lockheed parcels out production of the parts and subsystems in truly national fashion. In all, Lockheed boasts that one thousand suppliers in forty-two states play a role in equipping the F-22.

Lockheed has spread the wealth on other projects as well. In late December 2006, Lockheed divided up $376 million of work on a new Patriot missile system contract between facilities in Grand Prairie, Texas; Lufkin, Texas; Camden, Arkansas; Huntsville, Alabama; Chelmsford, Massachusetts; Clearwater, Florida; and Atlanta, Georgia. In early January 2007, the corporation divvied up $28.5 million of work on the Acoustic Rapid Commercial Off-the-Shelf Insertion sonar system among plants in Manassas, Virginia; Portsmouth, Rhode Island; Oldsmar, Florida; Chantilly, Virginia; Syracuse, New York; Chelmsford, Massachusetts; St. Louis, Missouri; and Houston, Texas.

The Chambliss Amendment was passed on the premise that a multiyear deal, which would lock the government in for bulk purchases of F-22s, would be a cost-saving mechanism—even though both the Government Accountability Office and the Congressional Research Service said otherwise. Chambliss claimed to have based his assessment on an "independent" analysis by the Institute for Defense Analyses (IDA), a federally funded nonprofit research center that focuses on technical aspects of national security issues. However, it was later revealed by the Defense Department's inspector general that the then head of the IDA, retired navy admiral Dennis C. Blair, had "violated conflict-of-interest rules when he failed to distance himself from two reports that could have affected companies in which he had a financial interest." Blair, it turned out, sat on the boards of two of those one thousand F-22 subcontractors, which made electronic components that were sold to other F-22 subcontractors.

While jobs in home districts and conflicts of interest surely played key roles in preserving the plane, the F-22 and other pork-laden weapons systems are regularly enabled by a basic practice that makes the "iron triangle" what it is: Washington-style pay-offs. As Matt Taibbi noted in *Rolling Stone,* "Chambliss' amend-ment passed 70–28, with wide bipartisan support. Most all of the senators who voted for the bill, including Democrats like Joe Lieberman, Chuck Schumer and Daniel Inouye, had received gen-erous campaign contributions from Lockheed Martin, the maker of the F-22, and from subcontractors like Pratt and Whitney." Such efforts are standard operating procedure for Lockheed, which spent more than $59 million on campaign contributions and lob-bying between 2000 and 2006, and—according to OpenSecrets.org, a Web site sponsored by the Center for Responsive Politics that analyzes filings from the Federal Election Commission—gave 59 percent of its money to Republicans and 41 percent to Democrats between 1990 and 2006.

But don't feel bad for the Democrats. The columnist Derrick Z. Jackson, writing in the *Boston Globe* in 2007, noted that among the ten senators who received the most money in campaign contribu-tions from defense contractors in the 2006 election cycle were the following six: Democrats Edward M. Kennedy of Massachusetts, Hillary Clinton of New York, Christopher Dodd of Connecticut, Dianne Feinstein of California, Bill Nelson of Florida, and Democrat-turned-independent Joe Lieberman of Connecticut, who "col-lected 60 percent of the $1.4 million the industry lavished among the top 10." For what did they deserve these riches? Clinton, notably, was found to have slipped twenty-six earmarks—line items (generally pork-barrel projects) inserted into huge spending bills to direct funds to a specific project without any review—into the defense budget. That was the second highest among senators. As for Kennedy, the *Globe* found that he "slid $100 million into the 2008 defense authorization bill for a General Electric fighter engine that the Air Force said it did not need."

The classic iron triangle—Congress, big military contractors (like Lockheed, Boeing, Northrop Grumman, the General Dynamics

Corporation, Raytheon, Halliburton, the Bechtel Corporation, and BAE Systems) and the Pentagon—has always formed the essential core of the military-industrial complex. These firms still reign supreme as the primary weapons-producing "merchants of death," but huge arms dealers, like Lockheed Martin and Boeing, are now only a portion of the story. While they may still rake in the largest single sums of any Pentagon contractors, their total take—Lockheed's was number one in 2006 with $26,619,693,002, or 9.02 percent of all contracts—is dwarfed by the combined totals of the rest of the DoD's contractors. These include big-name companies, small firms, and organizations you might never suspect of being on the military dole, ranging from Columbia TriStar Films and Twentieth Century Fox to Velda Farms ("a leader in the dairy industry for over 50 years") and from the National Vitamin Company of Porterville, California, to the American Meat Institute ("the nation's oldest and largest meat and poultry trade association") and the American Medical Association (which is dedicated to promoting "the betterment of public health"). These entities now form the bulk of the Complex, turning the iron triangle into a collection of "iron myriagons" (ten-thousand-sided polygons).

2

THE MILITARY-ACADEMIC COMPLEX

A few years after President Eisenhower pointed to the "unwarranted influence" of the military-industrial complex, Democratic senator J. William Fulbright of Arkansas spoke out against the militarization of academia, warning that, "in lending itself too much to the purposes of government, a university fails its higher purposes." He drew attention to the existence of what he called the military-industrial-academic complex, or what historian Stuart W. Leslie has since termed the "golden triangle" of "military agencies, the high technology industry, and research universities."

Defining and understanding this complex has never been simple. In actuality, the military-academic complex has two distinct branches. The first is the official, out-and-proud, but oft-ignored, melding of the military and academia. Since 1802, when Thomas Jefferson signed legislation establishing the U.S. Military Academy, the United States has been formally merging higher education and the art of warfare. The second is the increasingly militarized civilian university.

WAR-MAKING U

West Point, Annapolis, the Air Force Academy. The mere mention brings to mind a vision of dashing, broad-shouldered, square-jawed cadets in sharp uniforms (or perhaps the shadowy specter of rampant sexual harassment and rape); but when it comes to military education, if you're only considering the big-three service academies—even with the Merchant Marine Academy, the Coast Guard Academy, and private military schools like the Citadel thrown in for good measure—think again.

As it turns out, the Pentagon has an entire system of education and training institutions of its own, including the many schools of the National Defense University system: the National War College, the Industrial College of the Armed Forces, the School for National Security Executive Education, the Joint Forces Staff College, and the Information Resources Management College, as well as the Defense Acquisition University, the Joint Military Intelligence College—open only to "U.S. citizens in the armed forces and in federal civilian service who hold top secret/SCI (Sensitive Compartmented Information) clearances"—the Defense Language Institute Foreign Language Center, the Naval Postgraduate School, the Naval War College, Air University, the Air Force Institute of Technology, the Marine Corps University, and the Uniformed Services University of the Health Sciences, among others, totaling approximately 150 military-educational institutions.

While the service academies train tomorrow's military officers, the students enrolled in the National Defense University (NDU) are selected commissioned officers, with approximately twenty years of service, and civilian officials from various agencies, including the Department of Defense. They are schooled in a curriculum that emphasizes "the development and implementation of national security strategy and military strategy, mobilization, acquisition, management of resources, information and information technology for national security, and planning for joint and combined operations." Further, NDU is dedicated to promoting "understanding and teamwork among the Armed Forces and between those agencies of the Government and industry that contribute to national security." NDU also opens spots to "industry fellows" from the private sector, who, according to the former NDU president and air force lieutenant general Michael M. Dunn, "bring ideas from industry to the Defense Department."

JOE COLLEGE GETS DRAFTED

In 2002, NDU's budget topped out at $102.5 million. But while the formal military-academic complex of service academies and DoD institutions is a massive educational apparatus, its size, scope, and cost pale in comparison to those in the increasingly militarized civilian higher educational structure.

During World War II, as historian Roger Geiger noted, educational institutions carrying out weapons development received the largest government research and development contracts. Six of them, in particular—the Massachusetts Institute of Technology (MIT), the California Institute of Technology, Harvard University, Columbia University, the University of California at Berkeley, and Johns Hopkins University—received the then massive sums of more than $10 million each. (All were still on the DoD payroll in 2006.) Following the war, military entities like the Office of Naval Research (ONR) sought to establish, strengthen, and cultivate relationships with university researchers. By the time the ONR officially received legislative authorization to begin its work in August 1946, it had already entered into contracts for 602 academic projects employing over four thousand scientists and graduate students. Academia has never looked back.

For example, at the close of World War II, MIT was the nation's largest academic defense contractor. By 1962, physicist Alvin Weinberg sarcastically remarked that it was becoming difficult to figure out if MIT was a university connected to a multitude of government research laboratories or "a cluster of government research laboratories with a very good educational institution attached to it." By 1968, a year after Fulbright coined the phrase "military-industrial-academic complex," MIT already ranked fifty-fourth among all U.S. defense contractors. In 1969, its prime military contracts topped $100 million for the first time. By 2003, that number had grown to over $500 million, good enough to make

MIT the forty-eighth-largest defense contractor in the United States. And in 2005, the Cambridge, Massachusetts, school had crept up to forty-fourth place, pulling in over $600 million in DoD dollars.

Today, the scale of interconnections between military projects and academia is as dizzying as it is sweeping. According to a 2002 report by the Association of American Universities (AAU), almost 350 colleges and universities conduct Pentagon-funded research; universities receive more than 60 percent of defense basic research funding; and the DoD is the third-largest federal funder of university research (after the National Institutes of Health and the National Science Foundation).

The AAU further noted that the Department of Defense accounts for 60 percent of federal funding for university-based electrical engineering research, 55 percent for the computer sciences, 41 percent for metallurgy/materials engineering, and 33 percent for oceanography. When the DoD's budget for research and development skyrocketed to $66 billion for 2004—an increase of $7.6 billion over 2003, the results were almost immediate. In 2004, the AAU announced that the DoD had markedly increased its influence over both the electrical engineering and metallurgy fields, accounting for 68 percent and 50 percent of their federal funding, respectively. Not surprisingly, with this kind of clout, the Pentagon can often dictate the sort of research that gets undertaken (and the sort that doesn't).

The power of the Pentagon extends beyond an ability to frame or dictate research goals to significant parts of the civilian educational establishment. Higher education's dependence on federal dollars empowers the DoD to bend universities ever more easily to its will. For example, as Chalmers Johnson noted, until August 2002, Harvard Law School "managed to bar recruiters for the Judge Advocate General's Corps of the military because qualified students who wish to serve are rejected if they are openly gay, lesbian or bisexual." However, thanks to a reinterpretation of federal law, the Pentagon found itself able to threaten Harvard with a loss of all its federal university funding, some $300 million, if its law school denied access to military recruiters. Unable to fathom life ripped from the federal teat, Harvard caved, ushering in a new era of dwindling academic autonomy and growing military presence in the university.

But when it comes to education, the Department of Defense isn't mainly about the stick. It spends most of its time directing and channeling research by bestowing plenty of carrots, in the form of money and credentials (which, of course, lead to money). Take the National Security Agency (NSA), which runs the National Cryptologic School that "serves as a training resource for the entire Department of Defense." In addition to listening in on the globe and running its own school, the NSA doles out a seal of approval, in the form of a CAE designation ("Centers of Academic Excellence in Information Assurance Education") that puts schools in the running for lucrative DoD "Information Assurance Scholarship Program grant awards." In 2002, some thirty-six civilian schools earned CAE honors. In 2003, the list had expanded to fifty and included longtime DoD stalwarts like Stanford University, big state schools like the University of California at Davis and the University of Nebraska at Omaha, and lesser-known institutions like New Mexico Tech, West Virginia's James Madison University, and Vermont's Norwich University (the "oldest private military college in the United States"). In 2004, the Department of Homeland Security (DHS) joined the NSA as a sponsor of the program and the number of centers increased to fifty-nine. By 2006, there were some seventy-five such centers. And in 2007, there were "a total of 86 Centers across 34 states and the District of Columbia" from the University of Missouri-Rolla and Indiana University to Minnesota's Capella University and Our Lady of the Lake University in Texas.

In addition to the DHS, the NSA shares the spotlight with a host of other agencies and subagencies when it comes to the military-academic action. The credo of the Army Research Laboratory in Adelphi, Maryland, for instance, is "delivering science and technology solutions to the warfighter," which it strives to do by "put[ting] the best and brightest to work solving the [army's] problems." It employs "a variety of funding mechanisms to support and exploit programs at universities and industry." The Space and Naval Warfare Systems Command (SPAWAR) is also high on "university relationships" that provide it with "an excellent recruitment resource for high-caliber graduate and undergraduate students." Its SPAWAR Systems Center in Charleston, South Carolina, alone, has cooperative agreements with Clemson University, the University of South Carolina, the Citadel, the College of Charleston, Old Dominion University, North Carolina State University, Virginia Tech, Georgia Tech, the University of Central Florida, and North Carolina A & T State University.

MARCH (AND APRIL AND MAY AND . . .) MADNESS

Every year, the National Collegiate Athletic Association (NCAA) college hoops tourney has a Cinderella squad—a small-time five that shocks the field of sixty-five by knocking out a few top teams. In a military-academic-complex tournament such schools might come from the DoD's "Historically Black Colleges and Universities and Minority Institutions Infrastructure Support Program." These institutions don't get the big dollars of a national powerhouse, but they get modest awards to "enhance programs and capabilities at these minority institutions in scientific disciplines critical to national security and the DoD." Under this program, schools like Oglala Lakota College, Si Tanka University (chartered by the Cheyenne River Sioux Tribe), Sitting Bull College, and the College of Menominee Nation, among others, have been awarded grants, though these never seem to breach the million-dollar barrier.

Let's imagine a military-academic March madness tournament, with teams ranked by their slice of DoD Contract Awards for Research, Development, Test and Evaluation (RDT&E). There, a midmajor like the University of Dayton, which took in a respectable $22,273,173 in 2005, might beat up on Oglala Lakota College in the first round but would eventually fall to the might of the University of Hawaii ($34,400,055). Hawaii, however, would lose a squeaker to New Mexico State University's $36,305,495. Meanwhile, in another

bracket, the University of Texas would rout the University of Southern California by a score of $52,612,074 to $38,210,122. And
Carnegie Mellon would squeak by its interstate rival Penn State
$69,313,227 to $62,779,679.

Of course you wouldn't need to be a Las Vegas oddsmaker to
foresee which universities would make it out of the "Final Four."
For years, two schools have consistently been tops in RDT&E
money and have, year after year, duked it out for numero uno.
In 2002, Johns Hopkins University ($363,342,491) bested MIT
($354,932,746) by less than $9,000,000—the equivalent of a
three-pointer at the buzzer for these titans. The next year, it
wasn't even close as MIT raked in a whopping $512,112,618 to
Johns Hopkins's positively puny $300,303,097, making it the clear-
cut national champion! In 2004, Johns Hopkins's old college try
of $234,587,461 was again walloped by MIT's $604,950,277. And
in 2005, MIT completed the three-peat—a slam-dunk trouncing of
Johns Hopkins ($608,448,445 to $231,324,704), which cemented
its place in the annals of military-academic history.

In fact, MIT's numbers were good enough to rank it as eighth
on the DoD's 2005 list of the top one hundred recipients of RDT&E
money. But even that ranking doesn't convey the full dominance
of this champion. At number sixteen on the same list was the
MITRE Corporation, a not-for-profit originally made up of several
hundred MIT employees and formed in 1958 to create new technologies for the Department of Defense.

Today, MITRE provides engineering and technical services to
the federal government through three Federally Funded Research
and Development Centers (FFRDCs)—one of which, the DoD
Command, Control, Communications and Intelligence FFRDC,
happens to serve the Department of Defense. Moreover, MITRE,
itself, is thoroughly wrapped up in the military-academic complex. It provides support to a "broad base of customers within the
DoD and intelligence community," while "organizing and managing the first-of-its-kind Northeast Regional Research Center for the
Advanced Research and Development Activity," which includes,
among others, Brandeis University, Brown University, Columbia

University, Cornell University, Dartmouth College, Harvard University, Johns Hopkins University, the Massachusetts Institute of Technology, Princeton University, the State University of New York–Buffalo, the University of Massachusetts, the University of Pittsburgh, the University of Rochester, and Syracuse University.

With all this work for the DoD, MITRE brought in a cool $275,384,277 in RDT&E awards in 2005. If the funding dollars of MIT's offspring are added to MIT's total, the resulting $883,832,722 would move MIT out of the military-academic ranks and within striking distance of the military-corporate megagiant General Dynamics.

ACADEMIA'S UNNOTICED IDENTITY CRISIS

Today the military-academic complex is merely one of many readily perceptible, but largely ignored, examples of the increasing militarization of American society. The Pentagon has both the money and the muscle to alter the landscape of higher education, to manipulate research agendas, to change the course of curricula, and to force schools to play by its rules.

Moreover, the military research under way on college campuses across America has very real and dangerous implications for the future. It will enable or enhance imperial adventures in decades to come; it will lead to lethal new technologies to be wielded against peoples across the globe; it will feed a superpower arms race of one, only increasing the already vast military asymmetry between the United States and everyone else; it will make ever more heavily armed, technologically equipped, and "up-armored" U.S. war fighters ever less attractive adversaries and U.S. and allied civilians much more appealing soft targets for America's enemies. None of this, however, enters the realm of debate in the United States—which gives the idea of the ivory tower, or perhaps now an up-armored titanium tower, new meaning.

3

THE MILITARY-PETROLEUM COMPLEX

In November 2002, before the invasion of Iraq, then secretary of defense Donald Rumsfeld told Steve Kroft of CBS that U.S. saber rattling toward Iraq had "nothing to do with oil, literally nothing to do with oil." In 2003, Rumsfeld called the assertion that the United States had invaded Iraq to get at its oil "utter nonsense." ("We don't take our forces and go around the world and try to take other people's . . . resources, their oil. That's just not what the United States does.") In 2005, speaking to American troops in Fallujah, Rumsfeld reiterated the point: "The United States, as you all know better than any, did not come to Iraq for oil." Strong denials for sure, but were they true?

Rumsfeld's boss—and a man who knows a thing or two about addiction—President George W. Bush, proclaimed, in early 2006, that "America is addicted to oil." Later that year, Bush almost came clean about Iraq, admitting (after a fashion), according to Peter Baker of the *Washington Post,* that "the war is about oil." For the first time he used petroleum as a justification for continuing the occupation of Iraq, saying, "You can imagine a world in which these extremists and radicals got control of energy resources." Bush's acknowledgment was no great revelation. After all, oil is not only a key driver of the U.S. economy but also a major source of the nation's energy. As a former oilman (with Dick Cheney, the former head of oil-services giant Halliburton, as his vice president), Bush knew this all too well—hence an invasion of one of the Middle East's key oil lands topped by an occupation where, initially, looters were allowed to tear almost every part of the Iraqi capital to pieces, save for the Oil Ministry.

But Rumsfeld's military was more than just an armed occupier sent to lock down the planet's oil lands. It was also a known petrol addict. In his book *Blood and Oil,* Michael Klare laid out the little-acknowledged facts about the Pentagon's oil obsession:

> The American military relies more than that of any other nation on oil-powered ships, planes, helicopters, and armored vehicles to transport troops into battle and rain down weapons on its foes. Although the Pentagon may boast of its ever-advancing use of computers and other high-tech devices, the fighting machines that form the backbone of the U.S. military are entirely dependent on petroleum. Without an abundant and reliable supply of oil, the Department of Defense could neither rush its forces to distant battlefields nor keep them supplied once deployed there.

And the deployments DoD has "rushed its forces" to in recent years—in Afghanistan and Iraq—have sucked up massive quantities of oil. According to *Fuel Line,* the official newsletter of the Pentagon's fuel-buying component, the Defense Energy Support Center (DESC), from October 1, 2001, to August 9, 2004, the DESC supplied 1,897,272,714 gallons of jet fuel, alone, for military operations in Afghanistan. Similarly, in less than a year and a half, from March 19, 2003, to August 9, 2004, the DESC provided U.S. forces with 1,109,795,046 gallons of jet fuel for operations in Iraq. In 2005, Lana Hampton of the DoD's Defense Logistics Agency revealed that the military's aircraft, ships, and ground vehicles were guzzling 10 to 11 million barrels of fuel each month in Afghanistan, Iraq, and elsewhere. Yet, while the Pentagon reportedly burns through an astounding 365,000 barrels of oil every day (the equivalent of the entire nation of Sweden's daily consumption), Sohbet

Karbuz, an expert on global oil markets, estimates that the number is really closer to 500,000 barrels.

With such unconstrained consumption, recent U.S. wars have been a boon for big oil and have seen the Pentagon rise from the rank of hopeless addict to superjunkie. Prior to George Bush's Global War on Terror, the U.S. military admitted to guzzling 4.62 billion gallons of oil per year. With the Pentagon's post-9/11 wars and occupations, annual oil consumption has grown to an almost unfathomable 5.46 billion gallons, according to the Pentagon's possibly low-ball statistics.

As a result, the DoD had some of the planet's biggest petroleum dealers, and masters of the corporate universe, on its payroll. In 2005, alone, the Pentagon paid out more than $1.5 *billion* to BP PLC—the company formerly known as Anglo-Iranian Oil Company (on whose behalf the CIA and its British counterpart covertly overthrew the Iranian government back in 1953) and then British Petroleum. In 2005, the Pentagon also paid out over $1 *billion* to N. V. Koninklijke Nederlandsche Petroleum Maatschappij—also known as the Royal Dutch Petroleum Company (and best known in the United States for its Shell brand gasoline)—and in excess of $1 *billion* to oil titan ExxonMobil.

In 2005, ExxonMobil, Royal Dutch Petroleum, and BP ranked sixth, seventh, and eighth on the *Forbes* magazine's list of the world's five hundred largest corporations in terms of revenue. The next year, they bumped their way up to first, third, and fourth, respectively. They also ranked twenty-ninth, thirtieth, and thirty-first on the DoD's 2006 list of top contractors, collectively raking in over $3.5 *billion* from the Pentagon. The big three petrogiants are, however, only the tip of a massive, oily iceberg. Also on the Pentagon's 2006 list were such oil services, energy, and petroleum conglomerates as:

Ranking	Company name	Total take from the DoD (in dollars)
6	Halliburton	6,059,726,743
34	Kuwait Petroleum	1,011,270,194
45	Valero Energy	661,171,541

55	Refinery Associates of Texas	576,557,185
66	Abu Dhabi National Oil	494,286,000
70	Bahrain Petroleum	477,535,378
83	CS Caltex	356,313,452
94	Tesoro Petroleum	310,564,052

It's almost impossible to catalog all the companies with at least some ties to the oil game that are doing business with the Department of Defense, but if just the most obvious names on DoD's payroll are any indication, the U.S. military is mainlining petroleum from a remarkable assortment of places. For instance, in 2005 alone, the Pentagon payroll listed the following companies:

PENTAGON PETROL PUSHERS

A & M Oil	Carter Oil	Fannon Petroleum Services
Acorn Petroleum	CEL Oil Products	Farmers Union Oil
Advanced Petroleum Services	Chevron	Farstad Oil
Aegean Marine Petroleum	CITGO Petroleum	Frost Oil
AGE Refining	Colonial Oil Industries	Galp Energia
Al Mamoon Oilfield and Industrial Supplies	Colorado Petroleum Products	Gate Petroleum
Aloha Petroleum	Compañia Española de Petróleos	Gemini Petroleum NV
Alon Israel Fuel	ConocoPhillips	Gene Moeller Oil
American Petroleum Services	Cross Petroleum	Glenn Oil
Anderes Oil	Cosmo Oil Co	Gold Star Petroleum
Arguindegui Oil	D & W Oil	Golden Gate Petroleum
Armour Petroleum	Daigle Oil	Griffith Oil
Askar Petroleum	Dana Petroleum	Gulf Oil Limited Partnership
Bahrain Petroleum	Dime Oil	Haliburton's Energy Services Group
Baseview Petroleum	Drew Oil	Hanil Oil Refining
Big Bear Oil	Dunlap Oil	Harbor Petroleum
Black Oil	Ed Staub & Sons Petroleum	Harris Oil
Bosco Oil	Estacada Oil	Harry's Oil
BP	ExxonMobil	Haycock Petroleum

Hellenic Petroleum SA

Hyundai Oilbank

ICS Petroleum

Imperial Oil

Inlet Petroleum

International
 Oil Trading

Irving Oil

Jefferson City Oil

Jenkins Gas & Oil

John W. Stone
 Oil Distributors

Johnson Oil of Hallock

Kidd Oil

Kimbro Oil

Kuwait Petroleum

Lakeside Oil

Lamar Fuel Oil

Le Pier Oil

Lee Escher Oil

Main Brothers Oil

Mansfield Oil

McCartney Oil

McLure Oil

Merlin Petroleum

Morgan Oil

Motor Oil [Hellas]

Muddy Creek Oil
 and Gas

National Oilwell Varco

Nippon Oil

Northland Holding's
 Service Oil & Gas Inc

Navajo Refining

N.V. Koninklijke
 Nederlandsche

Odgers Petroleum

Oil Equipment Sales

Oil States Industries

Paramount Petroleum

Parkos Oil

Patriot Petroleum

Petro Air

Petro Alaska

Petro Star Valdez Inc.

Petrol Ofisi A.S.

Petroleos del Peru

Petroleum Management

Petroleum Partners

Petroleum Solutions

Petroleum Traders

Petrom S.A.

Pettit Oil

Phoenix Petroleum

Pitt Penn Oil

Potter Oil & Tire

Pro Petroleum

Rebel Oil

Repsol Petróleo S.A.

River City Petroleum

RKA Petroleum

Rogers Petroleum

RPL Oil Distributor LLC

Salathe Oil

SBK Oil Field Services

Seoil Gas

Shin Dae Han Oil
 Refining

Sinclair Oil Corporation

Shoreside Petroleum

South Pacific Petroleum

Southwest Georgia Oil

SPARK Petrol Ürünleri

St. Joe Petroleum

Strickland Oil

Sunglim Oil & Chemical

Superieur Petrol

Supreme Oil

T.A. Roberts Oil

Tate Oil

Tesoro Petroleum

Total S.A.

Tramp Oil and Marine
 and Tramp Oil
 Aviation

Transworld Oil Limited

Tri-Gas & Oil

U.S. Oil & Refining

Unocal Corporation

Valero Energy

Wallis Oil

Ward Oil

Western Petroleum

Western Refining

World Fuel Services

Wyandotte Tribal
 Petroleum

These 145 companies—far from constituting a complete list of energy-related firms on the DoD dole—took in more than 8 billion taxpayer dollars in 2005. To put that figure in perspective, that was more than the army paid out in the same year to the military-corporate powerhouses Lockheed Martin, Boeing, Northrop Grumman, General Electric, and the Bechtel Corporation, *combined*. Or over $2.7 billion more than it spent in 2005 on bombs, grenades, guided missiles, guided missile launchers, unmanned aerial vehicles, bulk explosives, all guns, rockets, rocket launchers, and helicopters.

No doubt due to his outfit's penchant for petroleum guzzling, in 2005, then secretary of defense Rumsfeld issued a memo calling on DoD staff to develop plans for employing alternative power sources and energy-saving technologies. As defense technology expert Noah Shachtman noted in early 2007, while the "Department of Defense might not care about the environment," it had met its green goals ahead of schedule. As a result, the Pentagon now touts itself as environmentally conscious, drawing attention to its use of wind power at the naval station at Guantánamo Bay, Cuba, and its dabblings in "cleaner, 'greener' hybrid fuel." On March 24, 2006, the Pentagon's American Forces Press Service published an article, "Hydrogen Fuel Cells May Help U.S. Military Cut Gas Usage," speculating that someday such technology might significantly reduce the military's "dependence on hydrocarbon-based fuels for transportation needs."

That day is not yet in sight. In fact, on March 23, 2006, the day before that article was published, the Pentagon quietly announced a series of DoD contracts that demonstrated the degree of its continuing addiction to oil: a $241,265,176 deal with Valero Energy; a $171,409,329 agreement with Shell Oil; separate contracts of $156,616,405 and $23,923,354 with ConocoPhillips; a $124,152,364 agreement with Refinery Associates of Texas; a $121,053,450 deal with Calumet Shreveport Fuels; a $118,374,201 jet fuel contract with Gary-Williams Energy Corporation; a $75,094,613 agreement with AGE Refining; a $43,994,360 deal with Tesoro Refining; and a $29,524,800 contract with Western Petroleum—all of which had a completion date of April 30, 2007.

Couple this with the fact that, on Rumsfeld's watch, the Environmental Protection Agency granted the DoD a "national security exemption" on trucks that failed to meet current emissions standards; that the army canceled plans to introduce "hybrid-diesel humvees" (the current military model gets just four miles per gallon in city driving and an equally dismal eight miles per gallon on the highway); and that it similarly dropped plans to retrofit the fuel-guzzling Abrams tank with a more efficient diesel engine (the current model, in service in Iraq, gets less than a mile per gallon), while the air force deep-sixed plans for the possible replacement of aging "surveillance, cargo and tanker aircraft engines"—and you're looking at a Pentagon patently incapable of altering its addiction-addled ways in any near future.

Since then, it's been more of the same. In March 2007, the Pentagon, now under Rumsfeld's replacement, Secretary of Defense Robert Gates, went on a two-day bender of epic proportions. On March 22 and 23, the DoD announced that it had struck "fixed price with economic price adjustment" deals, to be fulfilled by April 30, 2008, with ExxonMobil, Shell, ConocoPhillips, Valero, Refinery Associates of Texas, and ten other petrogiants to the tune of $4 billion. Another petro-binge occurred around the 2007 Labor Day holiday. Over the course of three days, the DoD acknowledged fuel contracts with BP, Chevron, Tesoro, and four others worth more than $1.4 billion.

The Pentagon needs two things to survive: war and oil. And it can't make the first if it doesn't have the second. In fact, the Pentagon's methods of mass destruction—fighters, bombers, tanks, Humvees, and other vehicles—burn 75 percent of the fuel used by the DoD. For example, B-52 bombers consume 47,000 gallons per mission over Afghanistan. But don't expect big oil (or even smaller petroplayers) to turn off the tap for peace. Such corporations are just as wedded to war as their most loyal junkie. After all, every time an F-16 fighter "kicks in its afterburners and blasts through the sound barrier," it burns through $300 worth of fuel a minute, while each of those B-52 missions means a $100,000 tax-funded payout.

According to retired lieutenant general Lawrence P. Farrell Jr., the president of the National Defense Industrial Association ("America's leading Defense Industry association promoting National Security"), the Pentagon is "the single largest consumer of petroleum fuels in the United States." In fact, it's the world's largest energy consumer, according to Shachtman. That, alone, guarantees the military-petroleum complex isn't going anywhere, anytime soon—just some fuel for thought next time you head out to a Shell, BP, Exxon, or Mobil station to fill 'er up.

4

GLOBAL LANDLORD

In 1790, the land area of the entire United States of America consisted of 864,746 square miles. By the year 2000, the nation spanned 3,537,438 square miles. In the intervening years, the United States also acquired an overseas empire.

In just three years, during the mid-nineteenth century, the United States annexed more than fifty Pacific islands. In all, the United States acquired title to at least one hundred noncontiguous island territories across the globe. And as with manifest destiny on the mainland, the U.S. military helped to drive the effort. For example, after attempting to purchase Midway Atoll in the nineteenth century, and being rebuffed by the Hawaiian king, the U.S. Navy simply seized control of the isle. The navy also helped to overthrow the Hawaiian government in 1893. Following its war with Spain, the United States acquired even more overseas holdings, including Guam, the Philippines, and Puerto Rico. In 1934, the navy took control of Johnston Atoll. The military eventually turned it into a storage site for chemical weapons.

During World War II, the United States seized the Marshall Islands and Micronesia, including Bikini Atoll, from Japan (which had taken control of the islands from defeated Germany, under the auspices of the League of Nations, after World War I). And in

1946, the U.S. military, which exercised total control over the islands, evicted the inhabitants of Bikini from their land and made it a nuclear test site (while an "out island" off the coast of Enewetak in the Marshalls eventually became a burial ground for radioactive waste). The nearby residents of Roi Namur were also thrown off their island to make way for a high-tech radar complex. In fact, back in 1984, the *Manchester Guardian* wrote that the Pentagon had "cleared the inhabitants off nearly every island in Kwajalein Atoll" and left all eight thousand of them crowded on the tiny, sixty-seven-acre Ebeye Island. Kwajalein is now home to the U.S. Army's hush-hush Ronald Reagan Ballistic Missile Defense Test Site.

In 2001, the *New York Times* also took a look at the ignored chain of islands that the U.S. military rules as a de facto colony (through a series of fifteen-year renewable agreements). According to the *Times,* "In the place of the simple fishing and farming existence they once knew on their lightly inhabited atoll . . . displaced Marshallese have been relocated to badly overcrowded islands like Ebeye and Enniburr, where cholera outbreaks are common and malnutrition is frequently reported."

GARRISONING THE GLOBE

In 2003, *Forbes* magazine revealed that the media mogul Ted Turner was America's top land baron, with a total of 1.8 million acres across the United States. The nation's ten largest landowners, *Forbes* reported, "own 10.6 million acres, or one out of every 217 acres in the country." Impressive as this total was, the Pentagon puts Turner and the entire pack of megalandlords to shame with more than 29 million acres in U.S. landholdings. Abroad, the Pentagon's "footprint" is also that of a giant. For example, the Department of Defense (DoD) controls 20 percent of the Japanese island of Okinawa and, according to *Stars and Stripes,* "owns about 25 percent of Guam." Mere land ownership, however, is only a fraction of the story.

In his 2004 book, *The Sorrows of Empire,* Chalmers Johnson

opened the world's eyes to the size of the Pentagon's global posses-
sions, noting that the DoD was deploying nearly 255,000 military
personnel at 725 bases in thirty-eight countries. Since then, the
total number of overseas bases has increased to at least 766 and,
according to a report by the Congressional Research Service, may
actually be as high as 850. Still, even these numbers don't begin to
capture the global sprawl of the organization that unabashedly
refers to itself as "one of the world's largest 'landlords.'"

The DoD's "real property portfolio," according to 2006 figures,
consists of a total of 3,731 sites. Over 20 percent of these sites are
located on more than 711,000 acres outside of the United States
and its territories. Yet even these numbers turn out to be a drastic
undercount. For example, while a 2005 Pentagon report listed U.S.
military sites from Antigua and Hong Kong to Kenya and Peru,
some countries with significant numbers of U.S. bases go entirely
unmentioned—Afghanistan and Iraq, for example.

In Iraq, alone, in mid-2005, U.S. forces were deployed at some
106 bases, from the massive Camp Victory, headquarters of the
U.S. high command, to small five-hundred-troop outposts in the
country's hinterlands. None of them made the Pentagon's list. Nor
was there any mention of bases in Jordan on that list—or in the
2001–2005 reports either. Yet that nation, as the military analyst
William Arkin pointed out, allowed the garrisoning of five thousand U.S. troops at various bases around the country during the
buildup to the war in Iraq. Furthermore, some seventy-six nations
have given the U.S. military access to airports and airfields—in
addition to who knows where else that the Pentagon forgot to
acknowledge or considers inappropriate for inclusion in its list.

Even without Jordan, Iraq, Afghanistan, and the over twenty
other nations that Arkin noted, in early 2004, were "secretly or
quietly providing bases and facilities," the available statistics do
offer a window into an organization bent on setting up franchises
across the globe. In fact, the Pentagon acknowledges thirty-nine
nations with an avowed U.S. base, has stationed personnel in over
140 countries around the world and boasts a physical plant that
consists of a total of at least 571,900 facilities. (Some Pentagon fig-

ures list an even higher number: 587,000 "buildings and structures.") Of these facilities, 61 percent are classified as buildings, 28 percent as structures, and 11 percent as utilities. Some 466,599 of these facilities are located in the United States or its territories. (The DoD owns or leases over 75 percent of all federal buildings in the United States.) The other 105,366 of these buildings, structures, and utilities are found abroad. All of these facilities, collectively, are valued at $658 billion.

According to 2006 figures, the army controls the lion's share of DoD land (52 percent), with the air force coming in second (33 percent), the Marine Corps (8 percent) and the navy (7 percent) bringing up the rear. The army is also tops in total number of sites (1,742) and total number of installations (1,659). But when it comes to "large installations," those whose value tops $1,584 billion, the army is trumped by the air force, which boasts forty-three megabases compared to the army's thirty-nine. The navy and marines possess *only* twenty-nine and ten, respectively. What the navy lacks in big bases of its own, however, it more than makes up for in borrowed foreign naval bases and ports—some 251 across the globe.

DIVERSIFICATION

Land and large installations, however, are not all that the Defense Department owns. The Pentagon also operates four Armed Forces Recreation Center (AFRC) resorts. Abroad there's the Edelweiss Lodge and Resort in Garmisch, Germany—an "idyllic location nestled at the foot of sweeping Alpine vistas" featuring multiple restaurants, an indoor pool, a "wellness center," an Internet lounge, video game rooms, a sports lodge, and a golf course, among other amenities. There's also the Dragon Hill Lodge in Seoul, South Korea, which boasts restaurants, lounges, game rooms, a health and fitness club (featuring a "luxurious redwood sauna, bubbling spa" and indoor lap pool), a specialty shopping mall, and an outdoor park with a "magnificent water fountain that dances to music." In the homeland, military folks can vacation at the Hale

Koa on Waikiki Beach in Honolulu, Hawaii, and the Shades of Green located in Walt Disney World, Orlando, Florida—whose Web site proclaims that it even "has its very own fleet of ships."

Until relatively recently, the U.S. Navy operated its own dairy, complete with its own herd of Holsteins. Even though it did get rid of its cows in 1998, the navy kept the 865-acre farm tract in Gambrills, Maryland, and now leases it to Horizon Organic Dairy. While it doesn't have a dairy, the army still operates stables—such as the John C. McKinney Memorial Stables, where many of the forty-four horses from its ceremonial Caisson Platoon live. It also has a big farm (the Large Animal Research Facility). In fact, the Pentagon actually owns hundreds of thousands of animals—from rats to dogs to monkeys. In addition to an unknown number of animals used for unexplained "other purposes," in 2001 alone, the DoD utilized 330,149 creatures for various types of experimentation.

Once known as the Post Exchange, or "PX," system, the Army and Air Force Exchange Service (AAFES) is a vast "joint military activity" that sells goods and services to "active duty, guard and reserve members, military retirees and their families" and then pumps the money back into the army and air force, including so-called morale, welfare, and recreation programs (like its AFRC resorts and recreation centers). The sheer size of this company-store operation is mind boggling. AAFES proclaims that it "operates more than 3,100 facilities worldwide, in more than 30 countries, five U.S. territories and 49 states. AAFES operates some 160 retail stores and 2,008 fast food restaurants, such as Taco Bell, Burger King, Popeyes and Cinnabon . . . [and] also provides military communities with convenience, specialty stores and movie theaters on installations worldwide, including locations in Operations Enduring and Iraqi Freedom." All that selling at all those locations translates into big bucks. From 2000 to 2004, AAFES revenues exceeded $37.9 billion.

The DoD also owns equipment, loads of it. For instance, it is the unlikely owner of "over 2,050 railcars, know[n] as the Defense Freight Rail Interchange Fleet." The DoD also reportedly ships 100,000 sea containers each year and spends $800 million annually on domestic cargo, primarily truck and rail, shipments. And when it comes to trucks, the army, alone, has a fleet of 12,700 Heavy Expanded Mobility Tactical Trucks (huge, eight-wheeled vehicles used to supply ammunition, petroleum, oils, and lubricants to other combat vehicles and weapons systems in the field) and 120,000 Humvees. All told, according to a 2006 Pentagon report, the DoD had a total of at least "280 ships, 14,000 aircraft, 900 strategic missiles, and 330,000 ground combat and tactical vehicles."

The Defense Logistics Agency (DLA), the DoD's largest combat support agency (with operations in 48 of the 50 states and 28 foreign countries), boasts, "If America's forces eat it, wear it, maintain equipment with it, or burn it as fuel . . . DLA probably provides it." Indeed, the DLA claims that it "manages" some 5.2 million items and maintains an inventory, in its Defense Distribution Depots (which stretch from Italy and Japan to Korea and Kuwait), valued at $94.1 billion. The DLA also runs the Defense National Stockpile Center (DNSC), which stores forty-two "strategic and critical materials"—from zinc, lead, cobalt, chromium, and mercury (over 9.7 million pounds of it in 2005) to precious metals such as platinum, palladium, and even industrial diamonds—at twenty locations across the United States. The DNSC, with a stockpile valued at more than $1.5 billion and $5.7 billion in sales of excess commodities since 1993, boasts, "There is no private sector company in the world that sells this wide range of commodities and materials."

All told, the Department of Defense owns up to having "over $1 trillion in assets [and] $1.6 trillion in liabilities." This is, no doubt, a gross underestimate given the DoD's historic penchant for flawed bookkeeping and the fact that, according to a study by its own inspector general, it cannot account for at least $1 trillion in money spent—or perhaps, according to former defense secretary Donald Rumsfeld, as much as $2.3 trillion. Cooking the books and stashing cash are fitting enough for an American organization, in the age of Enron, that thinks of itself not just as a government agency but, in its own words, as "America's oldest company, largest company, busiest company and most successful company." In fact, on its Web site, the DoD makes the point that it easily bests Wal-Mart, ExxonMobil, and General Motors in terms of budget and staff.

IT'S GOT THE WHOLE WORLD IN ITS HANDS

In addition to assembling a dizzying array of assets, from tungsten to tubas—in 2005, alone, it spent over $6 million on sheet music, music instruments, and accessories—the Pentagon also owns a great deal of housing—300,000 housing units worldwide. It is also a slumlord par excellence, with an admitted inventory of "180,000 inadequate family housing units." In fact, according to the Office of the Deputy Undersecretary of Defense (Installations and Environment), "Approximately 33 percent of all [military] families live on-base, in housing that is often dilapidated, too small, lacking in modern facilities—almost 49 percent (or 83,000 units) are substandard."

Meanwhile, the Department of Defense's own home, the Pentagon, bests the sultan of Brunei's Istana Nurul Iman palace, the largest private residence in the world—3,705,793 to 2,152,782 in square feet of occupiable space. The DoD brags that the Pentagon is "virtually a city in itself"—with 30 miles of access highways, 200 acres of lawn space (including a 5-acre center courtyard), 17.5 miles of corridors, 16 parking lots (with approximately 8,770 parking spaces), seven snack bars, two cafeterias, one dining room, a post office, a "credit union, travel agency, dental offices, ticket offices, blood donor center, housing referral office, and 30 other retail shops and services," a chapel, a heliport, and numerous libraries. Moreover, says the DoD, the construction of the Pentagon consumed a huge portion of the natural environment; its concrete reportedly contains "680,000 tons of sand and gravel from the nearby Potomac River."

The Pentagon's other properties are equally impressive. The combined worth of the world's two most expensive homes—the $138-million, 103-room "Updown Court" in Windlesham, Surrey, in the United Kingdom and Saudi prince Bandar bin Sultan's $135-

million Aspen, Colorado, ski lodge—don't even come close to the price tag on Ascension Auxiliary Airfield on a small island off the coast of St. Helena (the place of Napoleon Bonaparte's exile and death), which has an estimated replacement value of more than $337 million. Other high-priced facilities include Camp Ederle in Italy: $544 million. Morón Air Base in Spain: $621 million. Incirlik Air Base in Turkey: nearly $1.2 billion. Yongsan Garrison in South Korea: $1.3 billion. Thule Air Base in Greenland: $2.8 billion. Plus the U.S. Naval Air Station in Keflavik, Iceland, is appraised at $3.4 billion, and the various military facilities in Guam are valued at over $11 billion, in total.

The Pentagon's holdings—all 120,191 square kilometers—are almost exactly the size of North Korea (120,538 square kilometers). They are larger than any of the following nations: Liberia, Bulgaria, Guatemala, South Korea, Hungary, Portugal, Jordan, Kuwait, Israel, Denmark, Georgia, or Austria. While the 7,518 square kilometers of twenty microstates—the Vatican, Monaco, Nauru, Tuvalu, San Marino, Liechtenstein, Saint Kitts and Nevis, Maldives, Malta, Saint Vincent and the Grenadines, Barbados, Antigua and Barbuda, Seychelles, Andorra, Bahrain, Saint Lucia, Singapore, the Federated States of Micronesia, Kiribati and Tonga—*combined* pale in comparison to the 9,307 square kilometers of just one military base, the White Sands Missile Range in New Mexico.

DOWNSIZING?

In recent years, while it has been setting up hundreds of bases across the globe to support ongoing wars, the Pentagon has also been restructuring its forces in an effort to reduce troop levels at old Cold War megabases and close down less strategically useful sites. Does this mean less Pentagon control in the world? Don't bet on it. In fact, the U.S. military is exploring long-term options to dominate the world as never before.

The DoD, at the moment, maintains only a moving presence on the high seas. This may change. The Pentagon is now planning for future "sea-basing." No longer just a ship, a fleet, or "prepositioned materiel" stationed on the world's oceans, sea bases will be "a hybrid system-of-systems consisting of concepts of operations, ships, forces, offensive and defensive weapons, aircraft, communications and logistics." The notion of such bases is increasingly popular within the military due to the fact that they "will help to assure access to areas where U.S. military forces may be denied access to support [land] facilities." After all, as a report by the Defense Science Board pointed out "seabases are sovereign [and] not subject to alliance vagaries."

Imagine a future where the people of countries at odds with U.S. policies wake to find America's "massive seaborne platforms" floating just outside their territorial waters. It could be a world where the Pentagon might extend an even greater military reach. With the earth and the sea firmly in its grasp, the sky would be, quite literally, the limit for the DoD. Not surprisingly, then, in 2004, the "U.S. Air Force Transformation Flight Plan" outlined what, wrote Noah Shachtman, the editor of *Wired*'s "Danger Room" blog, "analysts call the most detailed picture since the end of the Cold War of the Pentagon's efforts to turn outer space into a battlefield . . . the report makes U.S. dominance of the heavens a top Pentagon priority in the new century." In fact, the U.S. military's outer-space policy statement proclaims, "Freedom of action in space is as important to the United States as air power and sea power."

PART II

TODAY'S CORPORATE BEDFELLOWS

With the basics of classic American militarism and militarization covered, set aside Lockheed, ExxonMobil, MIT, and all those global military bases for a moment to consider the range of today's military-corporate bedfellows. Sure, there are old, all-American favorites, like telecom giant AT&T, but there are also plenty of hip foreign companies too—like Toyota (more than $1.6 million from the DoD in 2006) and Volkswagen (over $1.9 million)—and younger U.S. civilian firms, like Google ($137,000-plus in 2006) as well as Starbucks and computer maker Apple, that you might never guess were between the sheets with the Pentagon. So put on your Oakley sunglasses, take a sip of that cup of designer coffee, and prepare for a wild ride into the deep-fried world of the military-doughnut complex (I kid you not!). This section offers snapshot views of how "civilian" firms have cozied up to the Pentagon and just how intertwined the civilian consumer world and military have become.

5

STARBUCKS GOES TO GITMO, iPODS IN iRAQ

When it comes to early twenty-first-century American culture, the Apple iPod and Starbucks coffee practically define the subject. In any U.S. city, what could be more ubiquitous than the sight of an individual in a state of mild caffeine withdrawal, disconnected from the world thanks to music delivered by Apple's white headphones, standing in a line at Starbucks waiting to order a high-priced Iced Quad Venti Breve Latte? Aside from their hallowed spots in the American cultural pantheon, these two thoroughly çivilian, all-American icons share something else: virtually unknown ties to the U.S. military.

CULTURE WARRIORS

Undoubtedly due to their iconic status, Apple and Starbucks were both drawn into the debate surrounding the Iraq War early in the conflict. With the invasion of 2003, movements in Asia and Europe arose to boycott American products in symbolic protest against the war. McDonald's, Coca-Cola, Budweiser, and, not surprisingly, Starbucks became favorite targets. For example, Reuters reported: "Sarah Stools, a 22-year-old German student of American studies, was headed for a Starbucks coffee shop in central Berlin [in March 2003] when her anti-war conscience got the best of her. 'I was thinking about going into Starbucks which I love, when I realized it was wrong,' she said. 'I'm backing the boycott because the war is totally unjustified.'" Others failed to see Starbucks's connection to the war, casting such protests as misguided. (Actually, a coffee-complex boycott would also have to spurn Kraft's Maxwell House, Nestlé's Taster's Choice, and Procter & Gamble's Millstone brand, the military's standard coffees of choice.)

Meanwhile, in the summer of 2004, the New York artist(s) "Copper Greene" unveiled her/his/their culture-jamming parody of the Apple iPod's ubiquitous grooving silhouette advertisements. ("Copper Green," according to Seymour Hersh, was the name of a Pentagon black ops program that "encouraged physical coercion and sexual humiliation of Iraqi prisoners in an effort to generate more intelligence about the growing insurgency in Iraq.") In the midst of outdoor arrays of multicolored posters touting the MP3 player's ability to put "10,000 Songs in Your Pocket," Copper Greene interspersed "iRaq" posters featuring one of the most iconic images of the Abu Ghraib torture scandal: the silhouette of a hooded Iraqi, balanced on a box, threatened with electrocution. Additionally, a bomb had replaced Apple's logo, and the tagline now read, "10,000 volts in your pocket, guilty or innocent."

Soon, similar posters began to appear in other cities. Across the Internet, on blogs, fotologs ("flogs"), and Web journals, viewers commented, offered critiques, and pontificated about the protest art. Many echoed a similar complaint. At one site, a critic wrote, "Since I live in NYC, I see the iPod ad everywhere. But to link torture in Iraq with using style of iPod ad doesn't make sense [sic]." Similarly, another blogger opined, "I'm not sure what rhetorical purpose is served by populating the popular iPod adverts with images of Iraq terror and war, except to make Apple somehow guilty of promoting the war by trying to sell a product."

STARBUCKS'S DOUBLE SHOT OF TORTURE

It's ironic that Apple got tagged with the taint of torture when Starbucks seems to be the operation that winds up in the places where U.S.-perpetrated torture is said to be occurring—and the company has apparently ignored all allegations.

In the past, Starbucks has sued individuals for copyright infringement involving even the use of parody logos. When a Starbucks coffeehouse clone opened in Ethiopia, the Seattle coffee giant was none too amused, stating, through a company spokesperson, "Even where it may seem playful, this type of misappropriation of a company's name (and reputation) is both derivative and dilutive of their trademark rights." But Valerie Hwang, a spokeswoman for Starbucks, proclaimed the company had no problem with a simulation Starbucks outlet set up by California National Guard troops serving in Afghanistan that used flavored syrups and donated beans sent twice a month from a Starbucks in California— even though this faux Starbucks was located at Bagram Air Base in Afghanistan, the site of notorious U.S. military and CIA high-security detention facilities from which horrific accounts of abuse, torture, even murder have emerged.

One-upping Bagram's combination fake-Starbucks and real-torture camp were the fully sanctioned Starbucks coffee kiosks that sprang up in early 2005 at "Camp America," the U.S. facility next door to the even more notorious prison camp at Guantánamo Bay, Cuba. Immediately upon arrival at the scandal-plagued site, Starbucks— whose mission statement speaks of "respect and dignity" and a commitment to "contribute positively" to local communities— began providing American troops with its signature "mochas, creamices and iced lattes," some 1,400 drinks per day. In 2006, the *Toronto Globe and Mail* reported that Gitmo was home to a total of "three Starbucks" serving up "decaf skim lattes for everyone but the detainees."

When questioned about its implicit support for the prison camp/torture center, in correspondence made available by the Business & Human Rights Resource Centre, Starbucks claimed it had always "been committed to operating its business in a socially responsible way and to living by a set of Guiding Principles that includes treating people with respect and dignity." It expected its "business partners to do the same." Further, Starbucks stated that the company would "refrain from taking a position on the legality of the detention center at Guantánamo Bay." Its Guantánamo stance was framed by a Starbucks spokesman as merely a response to popular will: "Many U.S. military personnel have let us know that they miss their Starbucks Experience while serving in remote locations and we are humbled that the troops frequently request Starbucks coffee. Many of our customers and partners (employees) also believe that it is important for Starbucks to support the men and women serving their country."

iPODS AND iRAQ

While Starbucks has, for years, donated coffee to U.S. military personnel stationed abroad, it was only in 2004 that the corporate coffee chain finally joined the ranks of Lockheed Martin and Boeing and became a full-fledged defense contractor, after inking a contract with the U.S. Navy. Apple, on the other hand, has had a long history of working with the U.S. military. For example, an April 1984 *Washington Post* article, "'Apples' Used to Aim Missiles," recounted the congressional testimony of Assistant Secretary of Defense Richard Perle, later an architect of the Iraq War, who revealed that the DoD had purchased fifty-five $25,000 Apple computers to aid military commanders in selecting targets for nuclear strikes.

Apple went on to supply the navy with Macintosh computers later in the decade and by the late 1990s had already inked its second cooperative research and development agreement (CRADA) with the Naval Air Warfare Center Weapons Division. During the first go-round, Apple assisted the navy with its High-Speed Anti-radiation Missile (HARM) Combat Maneuver Training, a ground-based training tool designed to familiarize F/A-18 pilots with the use of the aircraft's missile system. With the second CRADA, Apple America's senior vice president for sales, Robin Abrams, was not bashful about what she wanted from the collaboration. The computer manufacturer saw the "opportunity to play a pivotal role for

Apple" in attempts "to get more Apple computers and specific military computer applications into the military and government marketplace."

By the time of the Iraq War in 2003, Apple had certainly penetrated that marketplace—a fact that was likely unknown to the many bloggers who saw no connection between the U.S. occupation and the iPod maker. The targeting tool of choice for Major Shawn Weed, an intelligence planner with the army's Third Infantry Division as it readied for the invasion, was Apple's Titanium G4 PowerBook. Meanwhile, well-trained F/A-18 pilots went on to decimate the northwest portion of the city of Fallujah in April 2004; blasted away at areas south of Baghdad in November 2004; bombed Fallujah again in December 2004; and bombed the village of al-Bu Faraj in October 2005, killing at least fourteen civilians, according to witnesses.

As U.S. soldiers in Iraq carried iPods on missions and spliced them into the intercom systems in their vehicles, Apple continued to cash in on its Pentagon connection. In 2004, the company had no fewer than twenty separate contracts with the Department of Defense. By 2007, Apple's iPod had been fully drafted into military service when, at the behest of troops, language software developed to assist in the occupation of Iraq was adapted for the Apple device. The military even began using iPods in its recruiting efforts—as a lure to get kids to hand over personal information. And the National Guard began offering free iTunes music downloads, in exchange for the same type of information, on its Web site.

AS AMERICAN AS APPLE iPOD

According to nostalgic memory, an older version of American culture was typified by "Mom and apple pie." Today, Mom seems to have been dropped in favor of Starbucks's barista and apple pie traded in for Apple's iPod. But whatever the changes in American culture, war making has been a near-constant shaping influence.

So it's hardly surprising that the Caramel Macchiato from Starbucks that Americans love to slurp down and the Apple iPods that help them tune out, if not turn on, are part and parcel of the military-corporate complex. Nowadays, for U.S. icons such as these to be devoid of military ties would be downright un-American.

6

THESE BOOTS ARE
MADE FOR KILLING

The U.S. Army's Special Forces were formally created in 1952, but only during the Vietnam War did they became hot military properties. The "Green Berets" spawned books, toys, and even a John Wayne movie, before losing some of their luster when atrocities they had committed became front-page news. Out in the field, the Special Forces also regularly tested the Defense Department's newest, experimental equipment. One such innovation was the *footprint boot*. It sported a plastic-molded foot, not treads, on its sole and was designed to make a soldier's footprints indistinguishable from those of a barefoot farmer. The boots failed to fool anybody—and proved extremely difficult to walk in.

Today, the Special Forces have again become media darlings for their elite status and aggressive, globe-spanning style. Now, however, Special Ops troops leap into battle with boots fit for the runways of Paris and Milan. And why not? They have money to burn. Back in 2002, using the Global War on Terror and a mandate from Secretary of Defense Donald Rumsfeld as justification, air force general Charles Holland reported to the Department of Defense that his baby, the U.S. Special Operations Command, would need $23 billion in additional funds over five years (nearly doubling its previous allotment). Congress launched the new Special Forces era predictably—by awarding Special Ops forces $4.5 billion in

2004—almost $100 million more than its command had asked for and a nearly 50 percent increase over the previous year.

Perhaps sensing that, post-9/11, designer gear would be de rigueur, Oakley—best known as a purveyor of high-end, high-style sunglasses—formally announced, in 2003, that it had created "a boot designed specifically for the U.S. Elite Special Forces." The news had actually been leaked in 2002 when *Wired* magazine laid out the specs of the "21-ounce military boot that provides cushioning and stability for running and for parachute and fast-rope landings," but cautioned "weekend warriors—the [Standard Issue] Assault isn't for sale to civilians."

In 2003, armchair warriors were put at ease. They too would be able to slip on a "consumer version of the Elite Special Forces Standard-Issue Assault Boot" ($225) and an "Elite Special Forces Standard-Issue Assault Shoe" ($195). That spring, in fact, the footwear hit the shelves of stores like Los Angeles's superchic Fred Segal Feet, where the actor, comedian, and former TV alien Robin Williams snatched up the first pair.

The military version of the boot was produced by Oakley engineers in tandem with military specialists. "When they needed footwear for assault missions, they came to us," said Colin Baden, Oakley's president and chief designer. "The military saw Oakley innovation as a means to outperform their opponents." The Pentagon didn't have to look too far to see Oakley, which, since the mid-1980s, had been providing the U.S. military with products—most notably its signature merchandise: eyewear.

In 2003, Baden told CNN/Money, "We're hoping that the success of the U.S. military in Iraq helps sales of the products." Success for both parties, however, was elusive. As the *Wall Street Journal* headlined the matter in 2004: "Oakleys, Once Sports-Chic, Lose Their Eye for Fashion"—noting that sunglass sales "which represent more than half of the company's business—were down 6%." But it wasn't all bad news. According to the company, "Military industrial footwear increased significantly in 2004," and "U.S. military and other government sales grew 61.2 percent during 2004 to $27.1 million from $16.8 million in 2003."

Oakley sells eleven separate "families" of sunglass frames to the U.S. government. One of these, the Oakley SI M frame (military design)—black-framed sunglasses with a Plutonite ballistic lens that comes complete with a "cleaning bag" and storage case—is available through the Defense Logistics Agency for $90.40. The M frames also come in a laser eye-protection model, produced in conjunction with "Natick SOF [Special Operations Forces]—Special Projects, US army Special Operations Command, and Naval Special Warfare Development Group" with what else but "laser green" lenses.

Other Oakley products available only to the military—through vendor MAR-VEL International of Pennsauken, New Jersey—include:

XX. A model with Plutonite lenses and Unobtainium "nose-bombs" that "increase their grip with perspiration."

E Wire 2.1. Sunglasses with "XYZ Optics," which Oakley says allow for "razor-sharp clarity at all angles of view" and so are perfect for "S.W.A.T. teams and others who take their view of the world seriously."

Straight Jacket Eyewear. Another model with XYZ Optics as well as Unobtainium earsocks "that increas[e] grip with amphibious exposure."

In addition, the vendor Amrom International offers such Oakley-wear as:

Oakley SI Fives 2.0. The product description reads, "It's a solar war zone out there and all targets are fair game."

Oakley also provides the military with three different styles of goggles, as well as backpacks and assault gloves—a military model "developed from another [Oakley] glove platform," says Oakley's military liaison. By the looks of it, this item appears to be a sap glove—designed to pack an extra punch during hand-to-hand combat. But when asked, all an Oakley spokesperson would say was: "The knuckle is actually carbon fiber and it is to protect the hand from impact, whether that be from fighting with your hands or with weapons."

Getting straight answers from Oakley about most military matters is challenging, nor will Oakley representatives offer specifics on what the future holds. They have, however, hinted that they may be headed for boots geared for specific microclimates, such as marshy areas of Iraq. Right now, though, Oakley doesn't need new products to add to its millions in government sales. Its existing eyewear will do—at least as long as it is favored by the fickle fashion gods at the Pentagon.

Way back in 1994, the army adopted protective eyewear and in 1998 issued it to troops, but the general-issue sunglasses never took off, according to Sarah Morgan-Clyborne, who has been working on military eyewear for the army's Program Executive Office-Soldier for well over a decade. "We did not design a frame that was acceptable to soldiers," said Morgan-Clyborne. "Protection was important, but not a motivating factor." Colonel Chuck Adams, the senior optometry consultant for the army's Office of the Surgeon General, was more blunt. The old sunglasses were "a great product, but it doesn't look like an Oakley and doesn't look cool."

In July 2005, Adams announced that the army was embarking on fostering a "culture change" that would cause all soldiers to view sunglasses as don't-leave-camp-without-them gear. "We're talking about putting eyewear on half a million soldiers," Adams said. That's big money when sunglasses go for ninety dollars a piece, and, as Adams notes, Oakley's *cool* look is a hit with the troops.

At home, Oakley trades on a *badass* rep, cool-looking gear, and tough-sounding products like the "Killswitch" and "Crosshair" (men's footwear), the "Bullet" jacket and pants, the "Stomped" hoodie, "Strike" gloves, and watches dubbed "Detonator" and "Blade." Behind all the civilian martial and macho hype lies a deepening relationship with the military. Once you realize that, Oakley's "Bullet" takes on a new meaning.

7

A SNEAK PEEK AT THE MILITARY-TELECOM COMPLEX

In February 2006, amid disclosures of years of domestic spying by the military's supersecretive cryptologic service, the National Security Agency (NSA), *USA Today* reported that the agency had "secured the cooperation of large telecommunications companies, including AT&T, MCI and Sprint" to eavesdrop, despite legal prohibitions on such activities, on telephone and Internet traffic entering and leaving the United States. The report cited 2002 as the beginning of the period of secret NSA spying. A *New York Times* article, however, asserted that telecom companies had "for decades" assisted law enforcement and intelligence agencies in warrantless surveillance.

> The kind of cooperation now being provided by the companies has long been customary . . . American telephone executives have traditionally seen their companies' cooperation with intelligence agencies as a patriotic duty, industry officials say. The officials say they have not heard of companies' receiving regulatory breaks, contracts or other benefits for cooperation, both because a quid pro quo is not necessary and because it might expose the secret assistance.

At almost the same moment, Voice of America reported that the "Western Union Telegram Is No More." This seemingly unrelated news item is actually a key to understanding why "quid pro quo is not necessary."

Western Union's appearance in the news, even in tombstone form, was, in fact, quite fitting. Under Operation Shamrock—a program begun in the 1940s and taken over by the NSA after its founding in the 1952—the telegram giant dutifully delivered to the agency copies of each and every international message sent by Western Union from private citizens, businesses, and even foreign governments, in the United States, until the program was shut down in 1975. This long-standing corporate collusion in the military/intelligence spy game has indeed made the need for ordinary quid pro quo unnecessary, especially since such companies have always reaped "benefits."

Take, for instance, the cases of RCA Global and ITT World Communications—two major international carriers that joined Western Union in betraying their customers' privacy for over thirty years. In just one seven-year period in that span, from 1960 to 1967, RCA reaped $2 billion and ITT received $1.6 billion—remember this is 1960s dollars—in Pentagon contracts, nearly 20 percent of total sales for each company. Overall in that period, the telephone giant AT&T took in more than $4.1 billion in Department of Defense prime contracts. By 1968, it stood as the sixth-largest defense contractor (while RCA was twenty-sixth and ITT, twenty-ninth). By 2005, AT&T, no longer the same telephone titan, had tumbled down the list and ranked only as the eighty-ninth-largest Pentagon contractor, while rival MCI (then WorldCom) rated a respectable fifty-fifth place. Regardless, the monies at stake were still huge.

These contracts, not clandestine quid pro quo deals, ought to be viewed as the "benefits" that the telecoms received for "secret assistance" to the Pentagon in its maintenance of a national surveillance state. While far from complete, the following list of DoD contract announcements post-2002 is instructive.

April 4, 2002: MCI WorldCom is awarded a contract for the Defense Research Engineering Network (DREN), with a $450-million ceiling price over the ten-year period.

March 10, 2004: AT&T is awarded an "instant contract" worth $11,539,089, with "a total cumulative face value of $61,290,152."

April 23, 2004: MCI WorldCom is awarded a $7,879,000 contract for a modern, digital cellular, command-and-control system to link the various new sites of the Iraqi Armed Forces and the Coalition Military Assistance Training Team.

April 27, 2004: Sprint Communications is awarded a contract with the Defense Information Technology Contracting Organization, on behalf of the Department of Homeland Security, for $7,137,391, with a total cumulative face value of $19,620,776.

November 2, 2004: AT&T Wireless Service is awarded an estimated $20-million contract to provide nationwide commercial cellular phone service for the navy. This contract contains options, which if exercised will bring the total cumulative estimated value of the contract to $100 million.

December 16, 2004: MCI WorldCom is awarded two contracts for two circuits worth $15,400,082.

February 28, 2005: Sprint Communications is awarded an $8,704,107 firm-fixed price contract to provide a long-term lease of one OC-12 circuit between Misawa, Japan, and Andersen, Guam.

September 14, 2005: AT&T Government Solutions is awarded a $95-million contract to provide engineering and technical expertise to support the National Air and Space Intelligence Center Directorate of Data Exploitation Division.

January 24, 2006: AT&T Government Solutions is awarded an initial contract of $3 million, with an estimated life-cycle value of $95,397,395.

In 2005 alone, AT&T and MCI received a combined Pentagon payout of $807,669,962, while Sprint Communications took in at least $50,326,786. It's also worth noting that, in 2005–2006,

Verizon Communications, the nation's second-largest telecom company, received $277,204,960 from the Pentagon. So it was little surprise when, in October 2007, the *Washington Post* reported that Verizon admitted to congressional investigators that it had provided customers' telephone records to federal authorities "without court orders hundreds of times since 2005."

When you've been as deeply enmeshed in the military-corporate complex for as long as AT&T has, and you've got hundreds of millions of dollars in revenue at stake, as do Verizon, Sprint, and MCI as well, quid pro quo doesn't even enter the picture. Just look at ITT. While it got out of the telecom industry years ago, it's still a Pentagon favorite and took in over $2.5 billion in 2006, alone. In a reach-out-and-snoop-on-someone world, there's no need for favors when you're all one big happy family and surveillance is just a phone call away.

8

THE MILITARY-DOUGHNUT COMPLEX

According to the U.S. Army's 2004 *School Recruiting Program Handbook*, military recruiters are urged to ingratiate themselves in schools by delivering "donuts and coffee for the faculty once a month." Recruiters are also reportedly offering doughnuts to young people as a way of currying favor. Doughnuts may be a key asset in the military's battle plan for filling the ranks, but if you were to focus only on this aspect of the Pentagon's pastry of choice, you would be nowhere near the jelly center of the military-doughnut complex (MDC).

From the Pentagon to Guantánamo Bay to Iraq, from small-scale peddlers to industry giants (and even one of the largest military-corporate contractors), the DoD is well tied into the doughnut trade—more proof of the ridiculously expansive (if not simply ridiculous) reach of the military-corporate complex.

Still skeptical? Observe that back in 2001, the army spent more than $163,000 at just one doughnut emporium, Russell's Donuts, in Oklahoma. In 2002, the army spent $147,689 at Krispy Kreme Doughnut shops in Georgia and Kentucky (while dropping another $135,433 at Russell's). In March 2002, when the Navy Exchange Service Command began providing food service at the Pentagon, it promptly brought Dunkin' Donuts on board. That company's outlets can also be found on military installations from

the army's Fort Belvoir to Andrews Air Force Base to Naval Station
Norfolk. Dunkin' Donuts even shows up as a résumé builder.
According to the *New York Times,* senior military officials admitted
that almost none of the army interrogators stationed at the U.S.
naval base at Guantánamo Bay early in 2002 "had any substantial
background in terrorism, Al Qaeda or other relevant subjects."
One army intelligence reservist, however, did have impeccable credentials: He "had previously been managing a Dunkin' Donuts."
Perhaps he was pressed into double duty, as part of an interroga-
tion/percolation program, since during that same year, Dunkin'
Donuts donated five hundred pounds of coffee to the troops stationed at Gitmo.

Still need convincing that there's a MDC? Then consider that,
in 2003, Krispy Kreme Doughnuts franchises in Georgia and Kentucky took in $171,621 in DoD dollars, while one Krispy Kreme
franchise in Ontario Mills, California, donated 240 glazed (and,
after their long trip, possibly stale) doughnuts to marines stationed
in Iraq. Conceivably in a bid not to be outdone, Dunkin' Brands—
the parent company of Dunkin' Donuts—proclaimed that it "supports the brave men and women of the United States military,"
and to that end, holds "a monthly lottery to randomly select 50
United States military members . . . [to] receive a case of Dunkin'
Donuts coffee." Perhaps it should have also sent a few of their pas-
tries to Staff Sergeant Justin Hunt of the Rhode Island National
Guard's 173rd Long Range Surveillance Detachment. While on
patrol in Samarra, Iraq, in 2005, he was heard by a reporter to say,
"They got a [expletive] Dunkin' Donuts down in Qatar," before
continuing, "I'd kill for a doughnut."

Whatever doughnuts might be available in Qatar or Iraq, the
Hol-N-One Donut Shop on the U.S. Army base at Camp Arifjan,
Kuwait, is where, wrote one reporter, soldiers line "up 10-deep . . .
to get 'the best doughnut in town.'" One industry magazine even
suggested that another Kuwait base, Camp Doha, should be
renamed Camp Donut due to the presence of a twenty-thousand-
square-foot permanent facility operated by the Army Air Force

Exchange that "houses a concession that specializes in the deep-fried, punctured confections."

During 2005, alone, the DoD spent at least $1.2 million at the Donut House Company of Kuwait; nearly $500,000 at two Krispy Kreme outlets (one in Kentucky, the other in North Carolina); over $105,000 at Georgia's Augusta Doughnut Company; and more than $30,000 at a Tampa Bay Krispy Kreme that goes by the name Gulf-Fla Doughnut Corporation. In 2005, the military also paid more than $1.4 million to Super Bakery—a company owned by the former NFL great Franco Harris (of "Immaculate Reception" fame). The company's signature product is the Super Donut, a vitamin- and protein-enriched confection. Today, according to Ken Hoffman writing in the *Houston Chronicle*, these überpastries are served "at military bases around the world," including the mess tents of occupation troops in Iraq.

Perhaps sensing that doughnuts were a new hot growth field for the Complex, top Pentagon contractor, the Carlyle Group, joined two other firms, in late 2005, to buy up Dunkin' Brands and launched efforts "to expand the unit's strongest brand, Dunkin' Donuts, across the United States and internationally." With the purchase, 2004's twentieth-largest military contractor promptly became part owner of "the largest coffee and baked goods chain in the world . . . [with] more than 6,500 shops in 29 countries world-wide."

Whether it be doughnuts or bowling (as of 2004, the army had well over one hundred bowling alleys on bases worldwide, 1,830 lanes in all) or almost anything else of your choice, the case can be made for the existence of a discrete "-complex." This has less to do with the existence of actual doughnut (or bowling) industry ties with the Pentagon than with the immensity, deep pockets, and reach of a military establishment that stretches into every conceivable corner of economic life.

PART III

THE HIGH LIFE

Those corporate pals aren't the only ones having fun when it comes to the Complex. The military and the Central Intelligence Agency are also living it up on U.S. tax dollars—and blowing taxpayer money on perks, pork, and pillows (the kind that cost $1,800 each).

If you thought the days of the Pentagon shelling out $640 for toilet seats and $436 for an ordinary claw hammer, as it did in the high-flying 1980s, were over, you're absolutely, positively right. These days you need to add about $400 to the cost of one of those 1980s toilet seats if you expect to get a Pentagon-approved "nacho cheese warmer." And what used to buy a claw hammer, in the halcyon Age of Reagan, now barely covers the cost of that new necessity for military adventures in the Middle East—a decorative "genie lamp."

Given such expenditures, it's hardly surprising to find out that the Department of Defense has never actually undergone a financial audit. Speaking in 2006, Senator Tom Coburn (R-Oklahoma) made special note of the DoD's "inability to produce auditable financial statements":

> In other words, they can't undergo an audit, much less pass one. If DoD were a privately-owned company, it would have been bankrupt long, long ago. In 2004, the Department set the goal of undergoing a full audit by 2007. That deadline has not been met, and in fact, has been moved to the year 2016 . . . Americans are being asked to wait a full 10 years before their dollars are tracked well enough for the Department to fail an audit. And that seems to be the new objective of financial managers at DoD—to get to a place where DoD can actually fail an audit. Passing the audit is a pipedream for some future date beyond 2016.

This total lack of accountability has real consequences. When the Pentagon drops thousands of dollars on a "sumo wrestling suit" or the CIA indulges its penchant for living high on the hog at five-star hotels across Europe, it reflects one constant that has carried over from the old military-industrial complex: an intense commitment to wasteful spending. With that in mind, let's make like the jet set and take a no-expenses-spared tour of all the resorts, luxury hotels, and lavish living that the Complex has to offer.

9

HOW THE WASTE WAS WON

Back in the mid-1980s, the Department of Defense gained notoriety for outrageous spending: a $7,600 coffee pot, $9,600 Allen wrenches and—the most famous pork barrel item of them all—those $640 toilet seats. An outcry resulted and reform measures were enacted, yet the Pentagon continues to purchase high-priced pork with tax dollars, while the U.S. Congress adds to the squanderfest by tacking pet projects on to defense appropriations bills. As a result, one hand washes the other (and both feet) in a pork barrel that runneth over.

In 1996, though Ronald Reagan had long since passed from power, his DoD's legacy lived on—evidenced by the disclosure that the Pentagon doled out $2,187 for a door hinge for the C-17 cargo plane that should have cost $31. Similarly, in 1998, it was revealed that the Pentagon had spent "$714 each for 108 electrical bells previously priced at $47; $5.41 for each of 1,844 screw thread inserts, compared with a previous price of 29 cents; $1.24 for each of 31,108 springs previously priced at 5 cents; and $75.60 for each of 187 set screws previously priced at 57 cents."

In 2000, the U.S. General Accounting Office (GAO) found that the Space and Naval Warfare Systems Command (SPAWAR) had gone on a taxpayer-funded spending spree, buying loads of personal-use Palm Pilots (along with $100 designer carrying cases) and two $400 designer briefcases. Additionally, SPAWAR also used tax dollars to pay for Secretary's Day flowers. Meanwhile, Navy Public Works Center employees were buying so many high-end items, including "laptop computers, Palm Pilots, DVD players, an air conditioner, clothing, jewelry" as well as "eyeglasses, pet supplies, and pizza," that the GAO was unable even to tally the amount of money wasted.

The GAO reportedly also found that the DoD had purchased 68,000 first-class or business-class airline seats instead of coach tickets. And a separate GAO investigation discovered that, from 1997 to 2002, the Pentagon had wasted an estimated $100 million on unused airline tickets. But why fly commercial (even if you have the tickets) when, between 1999 and 2003, the Department of Defense purchased thirteen Cessna Citations (a business jet that then cost between $4.1 million and $7.6 million), two Gulfstream executive jets (each at up to $45 million), and two Boeing 737s (a VIP transport that costs $52 million) for a total of $272 million?

In their morning editions on September 11, 2001, before that day was seared into the American psyche, newspapers reported that Defense Secretary Donald Rumsfeld had vowed to battle Pentagon waste. "It could be said that it's a matter of life and death— every American's," said Rumsfeld. "Today, we declare war on bureaucracy." Instead, America went to war against an amorphous emotion ("terror"), and the Pentagon's long-standing spendthrift ways continued.

In 2002, for instance, the GAO found that the air force had purchased "a wide variety of home furnishing items . . . includ[ing] a sofa and armchair at $24,000, a designer table at $2,200, an executive pillow at nearly $1,800, and four leather chairs at over $1,000 each . . . [as well as] decorative accessories includ[ing] lithographs, wall hangings, decorative rock [sic] at nearly $19,000, and decorative knives costing hundreds of dollars each." Air force units deployed in Southwest Asia also spent $432 on a "genie lamp," $51,000 on cappuccino makers, a cool $1,000 on a "nacho cheese warmer," $3,395 on a sumo wrestling suit, almost $5,000 for cowboy hats, nearly $10,000 for Halloween decorations, more than

$36,000 on Palm Pilots, $276,000 for digital cameras, and almost $50,000 on a "bingo console." They even spent over $4,600 to buy "beach sand" for an air base in the United Arab Emirates surrounded by little else but sand.

Meanwhile, a 2003 GAO report laid bare the fact that "defense inventory systems [were] so lax that the U.S. army lost track of 56 airplanes, 32 tanks, and 36 Javelin missile command launchunits" and that the Pentagon could not account for $1 *trillion* in spending.

By 2005, little had changed—for the better, anyway. The Pentagon now accounted for a full 32 percent of federal programs on the GAO's list of those at "high risk due to their greater vulnerabilities to fraud, waste, abuse, and mismanagement." In fact, two of them, the DoD's inventory management and weapon systems acquisition systems, had been on the list every year since the inaugural report was issued in 1990. Commenting on the persistent problems, Comptroller General David M. Walker stated that the Pentagon's refusal to address such long-standing issues "results in billions of dollars in waste each year."

Not surprisingly, the GAO also discovered in 2005 that huge numbers of "new, unused, and excellent condition items were transferred and donated outside of DOD, sold for pennies on the dollar, or destroyed," while the Pentagon was buying identical items at a heavy loss to taxpayers. Out of $33 billion worth of matériel that the DoD classified as excess goods, $4 billion was found to be in excellent condition, but only about 12 percent of that was reused. For example, an army unit returned "172 pairs of new, unused extreme cold weather boots, which were subsequently sold for 40 cents per pair," while other units in need of boots purchased 214 identical pairs. Reutilization would have saved taxpayers $27,678.

It was also determined that the Pentagon was simultaneously selling unused biological- and chemical-weapons-resistant suits for $3 a piece, while buying hundreds of thousands of the same suits at $200 each. To make a point, GAO employees went out and purchased $2,898 worth of items on which the DoD had spent a whopping $79,649 and then promptly tossed out. By mid-2007, almost nothing, apparently, had changed, as the National Association of Aircraft and Communication Suppliers found that, between January and June of that year, at least $330 million worth of combat boots, helmets, vests, aircraft parts, and other gear was "being junked by the Pentagon rather than stored or sold as surplus to suppliers who sometimes sell it back to the military."

Like a spoiled child that needn't care for his toys because new ones are always on the way, the 2005 GAO report found that the Pentagon also left "excess property improperly stored outdoors for several months [where it] was damaged by wind, rain, and hurricanes." In one case, brand-new items were tossed into a trash Dumpster. Other things just went missing, including: "hundreds of military cold weather parkas and trousers and camouflage coats and trousers," 147 chemical and biological protective suits, 76 units of body armor, 47 wet-weather parkas, 21 pairs of chemical and biological protective gloves, 7 sleeping bags, as well as computer equipment, various other items—and, oh yeah, "5 guided missile warheads."

While the DoD was giving away, tossing out, and losing military matériel hand over fist, Bryan Bender of the *Boston Globe* noted that the Defense Appropriations Bill for fiscal year 2006 was filled with dubious *military* projects, such as: "$5 million to study mood disorders; $2.7 million to research a cancer vaccine; $4 million to find new ways to diagnose heart attacks; $4 million for something called the 'diabetes regeneration project.' None of them were included in the Pentagon's initial $363.7 billion spending request." In March 2006, it was also revealed that the Pentagon, by refusing to reuse electronic cargo shipping tags specifically designed to be used multiple times, had squandered an estimated $110 million since 1997—the worst of the waste occurring since the wars in Iraq and Afghanistan began.

In fact, the interminable war in Iraq has markedly exacerbated the Pentagon's already wild spending tendencies. By 2004, the GAO was reporting that billions of dollars were being wasted in Iraq. A significant percentage of this money was squandered by

Halliburton—the energy and base-building giant once headed by former Pentagon chief, and later vice president, Dick Cheney. In 2005, the Defense Contract Audit Agency found Halliburton had reaped $442 million in charges that were "unsupported" by documentation to justify the cost. The DCAA also found purchases to rival the Reagan years, including: "$152,000 in 'movie library costs,' a $1.5 million tailoring bill that auditors deemed higher than reasonable, more than $560,000 worth of heavy equipment that was considered unnecessary, and two multimillion-dollar transportation bills that appeared to overlap." Similarly, in 2007, it came to light that during the previous year the Pentagon paid another defense contractor "$998,798 in transportation costs for shipping two 19-cent washers." This was in addition to, according to the *Washington Post*, a "2004 order for a single $8.75 elbow pipe that was shipped for $445,640 . . . a $10.99 machine thread plug was shipped for $492,096 . . . [and] six machine screws worth a total of $59.94 were shipped at a cost of $403,463," in 2005. The pièce de résistance, however, was found in the testimony of the former food production manager at Halliburton's subsidiary Kellogg, Brown & Root (KBR), who told congresspersons that Halliburton charged the Department of Defense for as many as ten thousand meals a day it never served. At least the golden toilet seats and high-priced hammers actually existed!

10

LA DOLCE VITA
WAR ON TERROR

The Central Intelligence Agency has long loved the good life. In 1977, before the Senate Intelligence Committee, it was revealed that agents from a CIA safe house in San Francisco "infiltrated parties and dances to spike the drinks of guests with LSD and other hallucinogens and observe their freaked-out behavior." Other agents often were set up in genteel luxury. Robert Baer, a former CIA spook, fondly recalled his first agency assignment in Madras, India, this way: "It was a white, two-story stone and stucco house with a huge banyan tree and a pergola of jasmine that arched over the entire length of the driveway. Lined up under the veranda were my servants—all seven of them."

In 1996, three low-level CIA staffers were charged, according to a Department of Justice spokesman, "with stealing more than 108 credit cards intended for agents overseas and racking up $190,000 in bills for designer clothes, a 60-disc compact-disc system, a satellite dish, a 32-inch color television, automobile tires, tickets to Washington Bullets basketball games and $30,000 in cash advances from the cards—all in a nine-month period." This barely remembered scandal drew attention to the CIA's taste for lavish living. After all, for the better part of a year, charges for extravagant, big-ticket items and cash advances were racked up, apparently without raising either eyebrows or red flags. But the agency wasn't the only outfit around whose employees had a yen for living high on the hog.

Remember that, in September 2001, Donald Rumsfeld said, "In this building [the Pentagon] despite the air of scarce resources, money disappears," and he pointed out that the DoD was wasting "$3 billion to $4 billion" each year, while soldiers suffered. "The focus must remain completely on the warfighter," insisted the secretary of defense.

Despite the frugal rhetoric, a GAO report found that in 2001 and 2002, an estimated 72 percent of premium-class travel by the DoD was neither properly authorized nor justified. Apparently, the Pentagon "spent almost $124 million on over 68,000 airline tickets that included at least one leg of premium class service, primarily business class . . . based on [the GAO's] statistical sample, [it] estimated that senior civilian and military employees—including senior-level executives and presidential appointees with Senate confirmation—accounted for almost 50 percent of premium class travel." The report found that the DoD spent more on premium-class travel in two years than the total travel expenses—including airfare, lodging, and meals—of twelve major government agencies, including the Social Security Administration, the Departments of Energy, Education, Labor, Housing and Urban Development, and the National Aeronautics and Space Administration. A classic example of profligate Pentagon spending was revealed in the case of a DoD civilian employee who, along with three family members, "flew a combination of first and business class when they relocated from London to Honolulu. The travel order for the employee and his family did not authorize them to fly premium class, yet premium class tickets totaling almost $21,000 were issued, compared to an estimated cost of $2,500 for coach class tickets."

Commenting on such extravagant spending, enabled by travel cards issued by the Bank of America (which received more than $1 million from the DoD in 2006), Illinois representative Jan Schakowsky (D) said, "The irony is that these problems are occurring at the Department of Defense, an institution that places a premium on discipline, the chain of command and accountability. That makes the culture of waste, fraud and abuse that seems to permeate all aspects of DoD's fiscal operations all the more intolerable. This has to stop. It's unfair to our soldiers and to U.S. taxpayers."

The military's high-flying airfares were extravagant but paled in comparison to the Central Intelligence Agency's fight against frugality. According to Italian court documents, in December 2002, a CIA operative arrived in Milan and, eschewing a spartan safe house or low-profile lodgings, spent eleven days at the Milan Westin Palace, a posh hotel located in the "vibrant heart of Milan," which boasts "228 luxurious guest rooms—including 13 prestigious suites, 10 of which are endowed with a relaxant private Turkish bath"— and the Casanova Grill restaurant, located on the hotel's "enchanting private terrace." A gaggle of fellow CIA agents, eighteen in all, began arriving in January 2003 and, like their trailblazing counterpart, hit other posh digs, like the Milan Hilton ($340 a night) and the Star Hotel Rosa ($325 a night), where on room charges alone, the agency dropped $21,266.67 and $15,280.95, respectively.

According to the *Washington Post,* throughout that month, the covert agents were "regular patrons at the Hotel Principe di Savoia in Milan, which bills itself as 'one of the world's most luxuriously appointed hotels' and features a marble-lined spa and minibar Cokes that cost about $10." Seven of the agents stayed at the $450-per-night hotel for various lengths of time, accruing a bill of over $42,000 while "ringing up tabs of as much as $500 a day on Diners Club accounts created to match their recently forged identities." Another seven-person team, according to the *Chicago Tribune,* "spent $40,098 on room charges at the Westin Palace, a five-star hotel across the Piazza della Repubblica from the Principe, where a club sandwich is only $20."

In early February, the nineteen spies hit the Ligurian Riviera seaside resort town of La Spezia, checking into two hotels, but staying only a few hours. Some agents then made the three-hour drive back to Milan, while others headed to Florence for an overnight stay. Soon, at least thirteen of the agents were back in Milan, living it up at the Hilton, the Sheraton, the Galia, and the Principe di Savoia for another week at a cost of $144,984. Just after noon, on February 17, these budding 007s took to the Milan streets to find their Dr. No. He turned out to be a forty-two-year-old Egyptian-born cleric and suspected terrorist, Hassan Mustafa Osama Nasr, who, while walking to a mosque to attend daily prayers, was accosted by an eight-person CIA kidnap team. They promptly assaulted him with a chemical spray and threw him into a white van. The American abduction squad, carrying seventeen cell phones between them, began making phone calls to U.S. and Italian numbers, allowing Italian officials to track their movements as their vehicle traveled to the U.S. Air Force's Aviano Air Base.

A few hours later, Nasr was forced aboard a Learjet and whisked off to the U.S. air base at Ramstein, Germany. The cleric was then reportedly transferred to a Gulfstream IV executive jet—regularly leased by the CIA from Philip H. Morse, a part owner of the Boston Red Sox major-league baseball franchise (the Red Sox logo on the plane was covered for the "extraordinary rendition" flight)—and transported to Egypt, where the torture of prisoners is notoriously commonplace.

Nasr's *luxury* travel is par for the course when it comes to the CIA. In addition to the leased jets, the Special Collection Service, a joint project of the CIA and the National Security Agency, operates an entire fleet of luxury planes, generally Gulfstream and other executive jets, including a Boeing Business Jet (737), as well as regular military transports. Some of these executive jets are "registered to a series of dummy American corporations, such as Bayard Foreign Marketing, of Portland, Oregon," while others are owned by Premier Executive Transport Services, a "C.I.A.-linked shell compan[y]." They are operated by such concerns as Pegasus Technologies, Tepper Aviation, and Aero Contractors—a firm, apparently controlled by the CIA, that was founded by Jim Rhyne, a former chief pilot for Air America, the agency's Vietnam-era airline.

While a few members of the CIA team accompanied their captive to Germany, most stayed in Italy and continued living it up. At least four of the agents checked into luxury hotels in Venice, while others headed off for lavish digs in Florence, Tuscany, and the Italian Alps. A later Italian investigation of the kidnapping operation found that one of the operatives even owned a villa in the Piedmont region.

Not to be outdone, in 2003 and 2004, CIA agents involved in numerous "secret flights carrying detained or kidnapped Islamist terror suspects to interrogation centres and jails in Afghanistan, Egypt and elsewhere" lived the high life on the Spanish island of Mallorca, frequenting at least "two luxury hotels in Palma [de Mallorca]," where they reportedly "spent their spare time playing golf" and, after kidnapping Khaled el-Masri, a German car sales-man apparently mistaken for an Al Qaeda suspect with a similar name, in January 2004, racked up "a food bill of nearly $1,700 and an $85 bill for a massage."

One of the Mallorca hotels, the Gran Melia Victoria, prides itself on offering stunning sea views from its $1000-plus-per-night suites and is only a short distance from five local golf courses. The other hotel, the five-star Mallorca Marriott Son Antem Golf Resort & Spa, offers "36 holes of golf within walking distance from the hotel," five indoor and outdoor pools, a "Luxury Holistic Lifestyle Spa," as well as miniature golf, a mountain biking trail, tennis courts, and a sauna.

Could the DoD compete with such lavish living—especially in the wake of its new regulations that decreed, "Using Government funds to pay for premium-class travel (first and business) is strictly forbidden"? Since there was the caveat "except under certain cir-cumstances," signs pointed to *yes*.

In fact, the DoD's policy was a flight from accountability. Effec-tive November 17, 2003, officials in the Office of the Secretary of Defense were allowed to overrule written policy and approve first-class travel. But the SecDef's office wasn't alone. The loophole, a mile wide, was open to the executive secretaries of the Office of the Secretary of Defense and the defense agencies, the director of the Joint Staff, the secretaries of the military departments, who had the ability to further delegate authority to undersecretaries, service chiefs or their vice and/or deputy chiefs of staff, and four-star commanders or their three-star vice and/or deputy commanders.

Of course Rumsfeld didn't need to worry about travel restrictions, himself, since he flew in style on his own "highly modified Boeing 747-200 four-engine jet," known colloquially as "the Flying Penta-gon," that, according to the DoD's American Forces Information Service, "includes a group of flight attendants who prepare and serve food and beverages and much more." Meanwhile, according to a March 2004 memo, Rumsfeld's underlings were cleared to fly commercial airlines first class for six reasons or business class for nine reasons ranging from "security concerns" to the fact that the flight might last more than fourteen hours.

The DoD also remains competitive when it comes to swanky digs. Within the United States, the Pentagon maintains a list of "austere" hotels for the military traveler. These "flea-bags" include:

Hilton Miami Airport. Offers "tropical seclusion, exclusive accom-modations . . . and extraordinary dining . . . situated on a private peninsula in the heart of Blue Lagoon, a 100-acre freshwater lake . . . [with] jogging trail . . . a 3,000-square-foot swimming pool, and tennis and basketball courts."

St. Anthony: A Wyndham Historic Hotel. A San Antonio, Texas, landmark that boasts "timeless elegance and first-class service . . . lavish carpets, bronzes and works of art from around the world . . . [including] French Empire antiques."

Hilton Alexandria Mark Center Hotel. "Adjacent to a 43-acre botani-cal preserve [and] provides a tranquil oasis for travelers to Wash-ington [D.C.]. From the hotel's towering glass atrium and Italian marble clad lobby . . . guests know from the minute they arrive that they are assured a memorable stay . . . with the services and amenities that only a first-class hotel can provide."

The DoD also boasts its own Armed Forces Recreation Centers—special military-only resorts including the Hale Koa Hotel on Waikiki Beach in Hawaii. But a resort of its own in Waikiki is sim-ply not enough for the DoD. Its records show that, in 2004, the Pentagon sent U.S. tax dollars to the Hawaii Prince Hotel Waikiki, which boasts "spectacular all-oceanfront accommodations," including fifty-seven luxury suites, a golf course created by the legendary Arnold Palmer and the renowned course designer Ed Seay, tennis courts, a day spa, and a beauty salon, among other amenities. In 2005, the DoD spent over $29,000 at the Ocean View Hotel in Waikiki and in excess of $41,000 at the Hilton Hawaiian Village Beach Resort and Spa—a twenty-two-acre beachfront complex that contains "the widest stretch of white sand on Waikiki" and a beachfront "Super Pool" that is the largest on the island. That same year the DoD also paid more than $2.5 million to Waikiki Beach Condominiums and spent over $176,000 at the Doubletree Alana Hotel-Waikiki, "an intimate boutique hotel." In 2006, it was back for more, shelling out over $80,000 to the Doubletree Alana Hotel-Waikiki.

Military folks—at least of a certain rank or position—enjoy the good life when traveling overseas, even on missions. In 2004, an Associated Press story revealed that special agents of the Naval Criminal Investigative Service used "a five-star Kuwaiti hotel" as an interrogation center. Living high on the hog may be habit forming because, in 2005, the AP also revealed that "American military personnel . . . frequent Jordan's five-star hotels while on leave in Iraq." That same year, the *New York Times* exposed the fact that when two U.S. Special Forces snipers were arrested for gunrunning in Colombia, they were found "in a luxury gated community."

U.S. civilian contractors have adopted the same taste for living in high style as their military counterparts. Despite the fact that civilian contractors "have received battlefield commendations in Iraq" (the military later called these "mistakes" and rescinded the awards) and sometimes wear the same uniforms as U.S. troops, DoD travel regulations unequivocally state that "contractors are *NOT* Government employees." They just look, are treated, and act that way. After $1 million in taxpayer funds went missing, the Halliburton subsidiary KBR dispatched a "Tiger Team" to the Middle East to conduct an audit. According to Ed Harriman in the *London Review of Books,* "KBR's Tiger Team stayed at the five-star Kuwait

Kempinski Hotel, where its members ran up a bill of more than $1 million . . . the [U.S. A]rmy, whose troops were sleeping in tents at a cost of $1.39 a day . . . asked the Tiger Team to move into tents. It refused." And why wouldn't they say no? After all, those army tents lacked the "fruit baskets and pressed laundry delivered daily" offered at the Kempinski. And KBR folks are likely used to the high life, since previous concerns were raised "about the level of expenditure" of their employees at the Hilton Kuwait Resort, a "five-star beachfront hotel near Kuwait City." But if you think military folks are forced to forgo the Hilton's mile-long private beach, two pools, health club, tennis courts, spa, shopping arcade, and restaurants, for cheap tents—reconsider. DoD records show that the Pentagon paid more than $26,000 to the hotel in 2004, over $305,000 in 2005, and more than $284,000 in 2006.

PART IV

THAT'S ENTERTAINMENT

It seems only natural to move from the high life to the real glitz and glamour of the Complex—its ties to the entertainment industry. The military's been entertaining people for a long time and it still brings crowds to their feet with its spectacles. For example, between October 2005 and September 2006, the U.S. Army Field Band performed at 320 events, in 179 cities, in 45 states for more than 574,000 people. Meanwhile the Golden Knights, the army's skydiving team, reportedly entertains 12 million people a year at air shows and sporting events including Major League Baseball games, NASCAR races, and NFL football games.

Similarly, the Marine Corps band plays "500 public and official performances annually," while the air force reaches out with its popular Thunderbirds Air Demonstration Squadron, which according to the air force has been seen by more than "280 million people in all 50 states and 57 foreign countries." Counting television viewers, the numbers undoubtedly reach into the billions. The navy's aerial demonstration fighter jet squadron, the Blue Angels, has a similar performance record and, in 2005 alone, the navy's Leap Frogs parachute team played to big audiences at such diverse events as the San Diego Thunderboat Regatta, multiple Major League Baseball home openers, the Preakness Stakes horse race, and a Monday night football game.

Today, however, these classic techniques of the Complex have been augmented by much hipper methods—take, for one example, movies. While the U.S. military has long had a relationship with Hollywood, the ad hoc arrangements of old are over. Today, the air force operates airforcehollywood.af.mil, the official Web site of the U.S. Air Force Entertainment Liaison Office. The military has even set up a one-stop shop—on one floor of a Los Angeles office building—where the army, navy, air force, marines, coast guard, and Department of Defense itself have film liaison offices. Additionally, the DoD runs an entire "entertainment media division" from the Pentagon. But it doesn't stop there.

In 2007, the army, navy, and air force joined with Science Applications International Corporation (the tenth-largest DoD contractor in 2006, to the tune of $2.8 billion), the right-wing magazine *National Review,* and *In Hollywood* magazine, among others, to sponsor the first GI Film Festival—a three-day event with twenty-two film screenings, numerous panel discussions, and a "VIP cocktail reception including top military brass, filmmakers, Hollywood executives, and other opinion leaders from Washington DC." The festival caught something of the new interconnectedness typical of the Complex.

There's even a revolving door at work. For example, the director of the GI Film Festival, Major Laura Law, is not only a seventeen-year military veteran and a military intelligence officer in the Army Reserve but also the "the subject of a national recruitment media campaign for the Army reserve"—with a Web page devoted to her biography at the army's GoArmy.com. Or take Lieutenant Colonel Paul Sinor of the Office of the Chief of Public Affairs, the branch of the army that deals with the entertainment industry. Sinor actually served as a technical adviser to the Vietnam War–themed television series *Tour of Duty* in the 1980s. He retired from the military in 1991 and spent ten years in the movie industry—writing, acting, producing, and teaching screenwriting. Called back to the military in 2004, he traded in working on projects like his forgettable *Dead Men Can't Dance* (1997)—about a Special Ops team sent to destroy a North Korean nuclear facility—for Michael Bay's blockbuster *Transformers* (2007).

Eisenhower's military-industrial complex was olive-drab all the way—with the emphasis on drab. Today, the Complex is slick, *extreme,* and enmeshed in pop culture. And there's a good reason. By co-opting the civilian "culture of cool," the military-corporate complex is able to create positive associations with the armed forces, immerse the young in an alluring, militarized world of fun, and make interaction with the military second nature to today's

Americans. As such, the Complex is now interwoven in every aspect of American entertainment, from the video game industry to all forms of American car culture. So buckle up and enjoy the ride from Hollywood Boulevard to the Daytona International Speedway. Ain't it cool?

11

SIX BILLION MOVIES AND NO SEPARATION

In the late 1990s, Six Degrees of Kevin Bacon—a game in which the goal was to connect the actor Kevin Bacon to any other actor, living or dead, through films or television shows in no more than six steps—became something of a phenomenon. Spread via the Internet before becoming a board game and a book, Six Degrees has taken its place in America's pop culture pantheon among other favorite late-night drunken pursuits.

Here is a new variant of the game: the goal is to connect Kevin Bacon to the U.S. military. A commonsense approach would be first to consider Bacon's military roles—the ROTC cadet of his first feature film, the 1978 comedy classic *Animal House*, for example, or the Marine Corps prosecutor Captain Jack Ross in the 1992 film *A Few Good Men*. But the game isn't as easy as it looks. *Animal House* is hardly a promilitary project and the DoD actually denied *A Few Good Men* access to its facilities. The script, the Pentagon claimed, reinforced "the conclusion that not only is criminal harassment a commonplace and accepted practice within the Marine Corps, but that it requires a sister military service to uncover the wrongdoings and bring the perpetrators to justice." A spokesman for the film explained the Pentagon's decision: "It is certainly not a recruiting film," he admitted. So does that mean game over? Perish the thought. In reality, there are no degrees of separation between Bacon and the Pentagon because the actor had already starred in a military film—a real one. Bacon recalls: "After the [Vietnam] war was over in [19]75, I was already thinking about becoming an actor and I got sent out on this army recruiting film. It was a soft-sell kind of thing. I was a guy getting out of high school who didn't know what he wanted to do with his life, so I took the gig. It was my very first paying acting job."

As it happens, the Complex puts even Bacon to shame when it comes to connections in Tinseltown. The Pentagon might even be the ultimate Hollywood insider. So let's play a new version of the game Six Degrees of Kevin Bacon, with the military standing in for Kevin. The object is to follow a few of the thousands and thousands of linkages and connections between Hollywood and the military that have made the Department of Defense a genuine legend of the silver screen, from the Silent Era to the ramped-up military-movie complex of today, ending with—who else?—Kevin Bacon. Just sit back with a big bucket of popcorn and enjoy the show . . .

Let's go back to 1915, when in response to a request for assistance, U.S. secretary of war John Weeks ordered the army to provide every reasonable courtesy to D. W. Griffith's pro–Ku Klux Klan epic *Birth of a Nation*. The army came through with over one thousand cavalry troops and a military band. The film featured George Beranger, who would go on to star with Humphrey Bogart and Glen Cavender in *San Quentin* (1937)—in which a former army officer is hired to impose military discipline on the infamous California prison. Cavender had also appeared alongside actor/director Syd Chaplin, Charlie's brother, in *A Submarine Pirate* (1915), for which the U.S. Navy provided a submarine, a gunboat, and the use of the San Diego Navy Yard. (The film was even approved to be shown in navy recruiting stations.)

Syd Chaplin later starred in the nonmilitary *A Little Bit of Fluff* (1928) with Edmund Breon, who appeared in the 1930 World War I aviation epic *The Dawn Patrol*. That film was written by John Monk Saunders, who wrote another World War I drama, the Oscar-winning *Wings* (1927), featuring Gary Cooper. *Wings* received major support from the War Department (back in the days before it was called the Defense Department) and won the first Academy Award for Best Picture.

Gary Cooper provides the link to *Sergeant York*, a 1941 film directed by World War I Army Air Corps veteran (and *The Dawn Patrol* director) Howard Hawks, that was denounced by many as war-mongering propaganda. Hawks went on to direct actor Ray Montgomery in *Air Force* (1943), a Warner Brothers' film about a bomber crew serving in the Pacific, which received assistance from the Army Air Corps. In fact, the War Department even fast-tracked a review of the script because the film was deemed "a special Air Corps recruiting job."

That same year Montgomery also played a bit part, alongside Humphrey Bogart, in Warner Brothers' *Action in the North Atlantic* (with assistance from both the Merchant Marine and the navy). Bogart additionally starred with Lloyd Bridges, in Columbia Pictures' 1943 *Sahara,* a World War II epic made with the full cooperation of the U.S. Army. Bridges would go on to appear with both Van Johnson and Spencer Tracy in the nonmilitary *Plymouth Adventures* (1952). But long before that, both Johnson and Tracy took off in Metro-Goldwyn-Mayer's *Thirty Seconds Over Tokyo,* a film celebrating the 1942 "Doolittle Raid"—a U.S. terror-bombing effort that decimated civilian sites including factories, schools, and even a hospital—made, of course, with the assistance of the War Department.

Van Johnson fought his way through another MGM production, *Battleground* (1949), which not only featured tanks, trucks, and other gear loaned by the army, but, as extras, twenty members of the 101st Airborne Division. *Battleground* costarred John Hodiak, who, that same year, played alongside Jimmy Stewart in the World War II adventure film *Malaya*. Stewart had enlisted in the air force at the start of World War II, eventually becoming a colonel and earning the Air Medal, the Distinguished Flying Cross, the Croix de Guerre, and seven battle stars. He then served in the

Air Force Reserve and retired as a brigadier general. While in the reserves, he flew high in *Strategic Air Command* (1955), a film conceived at the urging of Curtis LeMay, the actual commander of the Air Force's real Strategic Air Command (SAC). Even with Cold War–era demands on its equipment, SAC provided Paramount with B-36 bombers, B-47 jet bombers, and a full colonel as a technical adviser.

But that was just one of SAC's (and LeMay's) connections to Hollywood. The 1963 film *A Gathering of Eagles,* for example, received SAC's wholehearted support. Written by *Battleground* screenwriter Robert Pirosh and featuring matinee idol Rock Hudson, it was praised for its realism by none other than LeMay. This realism can, in part, be explained by the fact that segments of the movie were shot on location at California's Beale Air Force Base and the filmmakers were granted access to SAC's underground command center in Omaha, Nebraska. Hudson (who had also appeared in 1948's air force–assisted *Fighter Squadron*), was even allowed to broadcast his lines over the SAC worldwide radio alert network.

Hudson later starred with John Wayne in *The Undefeated* (1969), but not before "the Duke" made his military-entertainment masterpiece *The Green Berets* (1968), which enjoyed the full backing of the Vietnam-embattled Department of Defense. On the say-so of President Lyndon B. Johnson's administration, the motion picture was allowed to shoot at the army's Fort Benning for 107 days. In addition, the Pentagon loaned airplanes, helicopters, weapons, troops, and advisers—a technical adviser to oversee matters of military assistance, a liaison at Fort Benning to arrange for equipment and men, and a liaison to oversee the progress of the production—to Wayne, the codirector and star.

With all that military input, *The Green Berets* proved to be, as *Variety* put it, a "whammo" and "boffo" box-office success. Critics, however, almost universally panned the motion picture. One *New York Times* film reviewer went so far as to call it "so unspeakable, so stupid, so rotten and false in every detail . . . vile and insane."

Denounced by antiwar protesters, who picketed theaters across the globe, the film was even attacked in the U.S. Congress after it was disclosed that the government received only $18,623.64 for the use of troops, equipment, land, and various services that might have cost taxpayers as much as $1 million. The army countered by suggesting that troops who worked on the film derived "training benefits" from their participation.

Wayne's *Green Berets* costar, George Takei (better known as Mr. Sulu on TV's *Star Trek*), was no stranger to the military-entertainment complex, having appeared in the 1960 Marine Corps–assisted *Hell to Eternity* and the 1963 film version of John F. Kennedy's *PT 109*. (For which the navy provided a destroyer, six other ships, various equipment, and a few sailors.) Takei (who would be "beamed up" in the navy-supported 1986 film *Star Trek IV: The Voyage Home*) also once starred with Grant Williams, an actor who later showed up in *Tora! Tora! Tora!*, Twentieth Century Fox's then unbelievably big-budget (at least $25 million) 1970 film. For that movie, the Department of Defense provided research assistance, stock footage, comments on multiple drafts of the script, a navy technical adviser, an airplane hangar that had been scheduled for demolition (which the film subsequently blew up), and use of navy ships at Pearl Harbor. Demonstrating a new willingness to go above and beyond for Hollywood, the navy even loaded thirty "Japanese" airplanes onto the aircraft carrier USS *Yorktown* for the attack.

Military-Tinseltown cooperation obviously goes back a long way, as far as the silent era. But in the 1970s there was a shift, producing a new, amped-up relationship, largely in response to a growing negative impression of the U.S. military engendered by the Vietnam War and the prospect of having to field an all-volunteer military. The military was hungry for help in rehabilitating its image— even lending support to "civilian" flicks—and the film industry was happy to oblige. Take Twentieth Century Fox's 1974 collaboration with the navy on the nonmilitary *The Towering Inferno* (1974). The navy lent helicopters, and the studio said thanks in the form of an acknowledgment in the credits. The film featured longtime military-entertainment stalwart William Holden, who had already appeared in *I Wanted Wings* (an army-aided 1941 propaganda flick that might as well have been a recruiting film) and *The Bridges at Toko-Ri* (made with navy assistance in 1955). He had also costarred in 1948's *Man From Colorado* with Glenn Ford—who acted alongside Charlton Heston in *Midway* (1976), a production that was allowed to use the USS *Lexington* aircraft carrier for two weeks of filming.

Heston, in turn, went on to star in *Gray Lady Down* (where he gives a scripted nod to defense giant General Dynamics). The 1978 submarine thriller benefited from the use of a real submarine, rescue ships, and sailors who served as extras, all courtesy of the navy. *Gray Lady Down* also featured actor Stacy Keach, who, in 1980, starred in the TV movie adaptation of Philip Caputo's *A Rumor of War*. Despite that film's focus on the atrocious side of the American war in Vietnam, the Marine Corps provided an adviser (who tempered some of the more disturbing portions of Caputo's memoir), the use of military facilities, and thirty marines.

Brian Dennehy, who also starred in *A Rumor of War*, would go on to act alongside Scott Glenn in the 1985 western *Silverado*. But before he became a cowboy, Glenn played the part of navy test pilot and NASA spaceman Alan B. Shepard in *The Right Stuff* (1983). That film, an adaptation of Tom Wolfe's book, was partially shot at California's Edwards Air Force Base and used various types of aircraft and equipment as well as air force personnel as extras.

Ed Harris, who blasted into orbit as astronaut John Glenn in *The Right Stuff* moved from the space capsule to the NASA control room in the 1995 blockbuster drama *Apollo 13* (air force extras and equipment loaned by Vandenberg Air Force Base, California). Beside him, in the copilot seat, was none other than Kevin Bacon. *Apollo 13* also featured Bill Paxton, who, a year earlier, had been seen in the Arnold Schwarzenegger blockbuster, *True Lies,* which benefited from Marine Corps assistance. Paxton had also acted in 1990's *Navy Seals* (helped by the navy) and, in 2000, would dive below the surface in the navy-supported submarine action-drama

U-571 alongside actor David Keith. In 2000, Keith played in the Robert De Niro/Cuba Gooding Jr. movie *Men of Honor* (made with the full cooperation of the navy). The next year Keith starred in *Behind Enemy Lines* (for which the navy and the Marine Corps, provided support, including the use of the USS *Carl Vinson* and Super Hornet fighter jets).

True Lies also provides another link in the military-entertainment matrix. The film's costar, Tom Arnold, had shared billing in *Exit Wounds* (2001) with Steven Seagal (whose 1992 film *Under Siege* and 1996's *Executive Decision* received navy and army cooperation, repectively) and Brue McGill, who also appeared in 2002's *The Sum of All Fears*. Shot on location at Whiteman Air Force Base and Offutt Air Force Base, *The Sum of All Fears* featured numerous USAF aircraft and enjoyed the input of multiple air force technical advisers. Another one of its stars, Morgan Freeman, also had a part in 1998's *Deep Impact* (helped by the army) and in *Shawshank Redemption* (1994), when he played opposite antiwar activist Tim Robbins, who provides a link to the navy-aided blockbuster *Top Gun*, starring Tom Cruise.

Freeman's costar in *The Sum of All Fears,* Ben Affleck, had a lead role in the 2001 historical drama *Pearl Harbor,* which was produced with the backing of the navy and even had its premiere on the deck of a nuclear-powered aircraft carrier. Affleck was joined in *Pearl Harbor* by Cuba Gooding Jr. (from *Men of Honor*), Tom Sizemore (from 1991's navy-aided *Flight of the Intruder*), and Josh Hartnett. That same year, Hartnett and Sizemore appeared in Ridley Scott's blockbuster *Black Hawk Down,* made with the full cooperation of the army. The Pentagon sent the film eight combat helicopters, 100 soldiers, including members of the 160th Special Operations Aviation Regiment, and two technical advisers.

Tom Sizemore, in turn, acted in the Vietnam War film *We Were Soldiers* (2002), for which much assistance was given by the army. One of his costars, Steven Ford, had appeared in *Armageddon* (1998), which was filmed, in part, at Edwards Air Force Base and Patrick Air Force Base, made use of USAF Thunderbirds aircraft, and employed Air Force personnel as extras and security personnel. The film also boasted a cast of big military-entertainment guns, including: Owen Wilson (from *Behind Enemy Lines*), Michael Clarke Duncan (who did voice-over work for the navy-aided SOCOM II: U.S. Navy SEALS video game), and William Fichtner (of *Pearl Harbor* and *Black Hawk Down*), while the movie's biggest gun was the now-familiar Ben Affleck.

Another costar in *Pearl Harbor* was Tom Everett, who also appeared in *Air Force One* (1997), starring Harrison Ford. The latter movie utilized USAF aircraft, air force personnel as extras and was filmed, in part, at both the Rickenbacker and Channel Islands Air National Guard bases. Its director, Wolfgang Petersen, also directed the George Clooney/Mark Wahlberg weather drama *The Perfect Storm,* which was partially filmed at the Channel Islands base, too, and involved air force aircraft, pilots, ground crews, and pararescue personnel.

Wahlberg also had a bit part in the 1994 Danny DeVito comedy *Renaissance Man* (made with army involvement). In fact, the Oscar-winning *Forrest Gump* received only limited help from the army, in part because *Renaissance Man* and another 1994 comedy, *In the Army Now,* starring Pauly Shore and David Alan Grier, sucked up so much of the military attention that year. Grier went on to appear in *The Woodsman* (2004) with Benjamin Bratt, who had previously been cast in the 1994 army-aided thriller *Clear and Present Danger* and would star in the ABC TV series *E-Ring,* a self-proclaimed "pulsating drama set inside the nation's ultimate fortress: the Pentagon," whose producer and cocreator Ken Robinson had worked in the actual Pentagon over "a couple decades." In case Bratt's military-entertainment bona fides needed further enhancement, it's worth pointing out that at his side in the nonmilitary *The Woodsman* was not only Grier but—you guessed it: Kevin Bacon.

In fact, one could take many (if not all) of Bacon's nonmilitary roles and quickly find connections that lead directly to the Pentagon. For instance, have a look at Bacon's distinctly unmilitary *Wild Things* (1998) and you'll find movie veteran Robert Wagner, who was featured not only in such navy-supported fare as *The Frogmen*

(1951), *The Towering Inferno* (1974), and *Midway* (1976), but also in the Marine Corps–aided *Halls of Montezuma* (1950), *Stars and Stripes Forever* (1952), and *In Love and War* (1958); the army-assisted *Between Heaven and Hell* (1956); the air force–supported *The Hunters* (1958); and finally in *The Longest Day* (1962), an epic about the D-Day landings made with the cooperation of the full complement of army, navy, *and* Marine Corps.

The point is, when it comes to military-entertainment connections, Bacon isn't special. As the Bacon/Wagner example demonstrates, almost any current actor—from Gwyneth Paltrow (in 2008's air force–aided *Iron Man*) to young actress Dakota Fanning (at the side of top-gunner Tom Cruise in the army-aided, Steven Spielberg–directed 2005 remake of *War of the Worlds*)—could be linked to the military by six degrees, if not many fewer. The reasons are simple. As David Robb, the author of *Operation Hollywood: How the Pentagon Shapes and Censors the Movies,* observed:

> Hollywood and the Pentagon have . . . a collaboration that works well for both sides. Hollywood producers get what they want—access to billions of dollars worth of military hardware and equipment—tanks, jet fighters, nuclear submarines and aircraft carriers—and the military gets what it wants—films that portray the military in a positive light; films that help the services in their recruiting efforts.

But recruiting is just part of the equation, and the phrase, "a positive light" is even a little soft. At the movies, the military gets sold as heroic, admirable, and morally correct. Often, it can literally do no wrong.

Speaking about the big-budget, live-action blockbuster *Transformers* (2007), Ian Bryce, one of the producers, characterized the relationship this way, "Without the superb military support we've gotten . . . it would be an entirely different-looking film . . . Once you get Pentagon approval, you've created a win-win situation. We want to cooperate with the Pentagon to show them off in the most positive light, and the Pentagon likewise wants to give us the resources to be able to do that." On the military side, air force master sergeant Larry Belen spoke of similar motivations for aiding the production of *Iron Man*: "I want people to walk away from this movie with a really good impression of the Air Force, like they got about the Navy seeing *Top Gun*," he said. But air force captain Christian Hodge, the Defense Department's project officer for *Iron Man*, may have said it best when he unabashedly opined, "The Air Force is going to come off looking like rock stars."

12

A VIRTUAL WORLD OF WAR

I'm breathing hard as I double-time it down a rubble-strewn street, hugging the outside wall of a building into which I just poured a heavy volume of rifle fire. I glance at windows and balconies for hidden snipers and holler at my fire team to move it. As we pile through a side entrance, I listen . . . voices . . . and it ain't English. I spot a circuit breaker box on the wall, take aim with my M-16A4 .233-caliber automatic rifle, and pump one round into it. Almost immediately, the room goes dark and a voice upstairs starts shouting. Was it Arabic? What do they speak here anyway? No matter. My team and I put on our night-vision goggles, proceed to the upper floors, and come upon a hallway lined with doors. I hear feet shuffling behind the first one and whisper orders to my team to execute a room takedown. We stack in front of the door, check our weapons, get ready, and bust in. Immediately, all hell breaks loose. I see a figure in the shadows and let loose with my M-16, cutting him down. To my right, my SAW gunner opens up on another man with his M-249 light machine gun. My official manual says that this high-powered weapon can be "devastating" to the enemy's morale. That guy's morale is the least of his worries now.

The room is clear, and I'm about to take an AK-74 automatic rifle (a later-model version of the AK-47 assault rifle) from one of the corpses when I hear a bullet rip right past me. Out of the corner of my eye I spot a muzzle flash as another shot narrowly misses a fellow team member. The shots are coming from the building across the street. A sniper! I take cover and try to zero in on him, but he's well protected and increasing his rate of fire. "OPFOR has me pinned," I roar—using military slang for *opposing forces,* or *bad guys.* Then more OPFOR start firing into the room, from the hallway. I guess shooting up their building, knocking out the power, and killing their comrades must have made them angry. I leave my team members to deal with them as I load my single-shot, rifle-mounted M-203 grenade launcher, edge back to the window, take hasty aim at the sniper, and fire a 40 mm grenade. The explosion is loud and blows out part of the wall of the building across the street. I wait and watch. Silence. I must have got him. Or at least chased him off.

In the meantime, my team has made short work of a bunch of bad guys, dead on the hallway floor. I proceed to take an AK-74 from one and ammo from another. Over the next hours I lead my team through the warrens of a thoroughly bombed-out Beirut—streets laden with concrete rubble, burned-out cars, and disabled armored vehicles. We take out machine-gun nests set up in storefronts, call in air strikes (in the heart of the city) by Cobra gunships, slog through ancient sewer systems, and engage in firefights—with fatigue-clad militiamen, turban-wearing "radicals," and occasional Syrian troops and Iranian Special Forces soldiers—in marketplaces, bombed-out bookstores, and restaurants. Suddenly, we're told that a high-value target is located in our area: Akhbar al'Soud, a Lebanese militia leader. We track him to an almost totally decimated building and fight our way inside, spending ammo with abandon and mowing down one local fighter after another, until we reach al'Soud himself, cowering alone on an upper floor. His hands are clasped behind his back, and he's willing, it seems, to be taken into custody. I walk up to him, raise my rifle, and butt-stroke him in the head. He crumples to the ground as my radio crackles that we've completed the operation. Maybe it was the adrenaline, but soon after, as we're appraised of the results of our mission, I discover that I killed al'Soud—the high-value target we were sent to find—with my unnecessary blow. Regardless, the operation is classified a success.

Almost immediately, a cable news network announces word of our triumphant mission, merely mentioning that Akhbar al'Soud is "no longer a factor" in the conflict. Only my team knows that I killed the man after the fighting was over. Only they witnessed me smash my rifle butt into his skull. And they know as well as I do that if I kill an enemy who has surrendered, the mission is supposed to be deemed a failure. But that didn't happen here. Nobody talked. And nobody even thought to mention the many civilian-occupied buildings I fired grenades into, raked with rifle fire, or targeted for mortar strikes. In any event, instead of being thrown into the brig, rousing music is played in my honor and I'm briefed on our next mission. I'm still in command. No one seems to give a damn about my war crimes.

It's true. I'm a murderer, a war criminal, and a U.S. marine. At least, in the digital world I am. It was through the Microsoft Xbox game Close Combat: First to Fight (2005) that I was *sent*, as real U.S. marines were in the 1950s and the 1980s, into Lebanon. Set in 2006, this act of digital Middle Eastern meddling, according to the game's story line, is the result of an American decision to impose its will after "several groups of insurgents" took over sections of the Lebanese capital, Beirut—the "largest, best-organized and best-funded" of them being "the radical Atash movement, led by Tarik Qadan, a local religious zealot of considerable influence."

My digital outfit, the 28th Marine Expeditionary Unit (Special Operations Capable), is a part of America's expeditionary invasion force, or as the game's promotional literature puts it, "the United States' 911 shock troops"—the "most feared shock troops on the planet." Of course when you're the "most feared" force on earth you do what you want and kill who you want—as I did in virtual Beirut and as the real marines have long done across the globe. (On-line background material for the game brags that the marines have been set "free of U.N mandated ROEs [rules of engagement] that guaranteed earlier failure.")

What is most noteworthy about Close Combat: First to Fight,

however, is not its military theme nor its realism. Instead, it's the way it came into being. The game is typical of a recently emerging trend that has melded the video game industry (and entertainment industries more broadly) with the U.S. military in a set of symbiotic relationships that literally immerse civilian gamers in a virtual world of war while training soldiers using the hottest gaming technology available. It's the creation of a digital cradle-to-grave concept in which games created by or for the military are used as recruiting tools and also, as it were, to pretrain youngsters. Then, when they're old enough to enlist, these kids find themselves using video game–like controllers to pilot real military vehicles and are taught tactics and trained in strategy using specially designed video games and commercially available, off-the-shelf games that have been *drafted* into service by the military.

In the late 1990s, the otherwise dreadful soundtrack for *Godzilla,* that blockbuster-flop of a movie, featured one cut that transcended its origins. "No Shelter," by rebel rap/rockers Rage Against the Machine, trashed both the movie ("And Godzilla pure mutha-fuckin filler, To keep ya eyes off the real killer") and a consumer-driven militarized Hollywood, writ large. "From the theaters to malls on every shore," the group decried: "Tha thin line between entertainment and war."

The line had by then grown thin indeed. Today, it hardly exists. The military is now in the midst of a full-scale occupation of the entertainment industry, conducted with far more skill (and enthusiasm on the part of the occupied) than America's debacle in Iraq. Consider the genealogy of Close Combat: First to Fight. Originally a training tool, called First to Fight, the software was developed for the U.S. Marine Corps by civilian contractor Destineer Studios (a video game developer and publisher that made training simulators for the CIA before the agency's own venture capital firm, In-Q-Tel, purchased an equity stake in the company). But the game wasn't strictly a civilian venture; it "was created under the direction of more than 40 active-duty Marines, fresh from the frontlines of combat in the Middle East . . . [who] worked side-by-side with the development team to put the exact tactics they used in combat into First to Fight." That civilian-created, military-aided training tool was then recycled into a civilian first-person shooter, rated "T" for "teen," with a marine on the game's packaging and a blurb that exclaims, "Based on a training tool developed for the United States Marines."

PLAY ALL THAT YOU CAN PLAY

Close Combat: First to Fight was hardly the first game to blur the civilian/military gaming divide (and certainly won't be the last). In 2002, the army launched America's Army (AA), a training and combat—the army balks at the term *shooter*—video game that was made available online and at recruiting stations free of charge. The game was the brain child of Lieutenant Colonel Casey Wardynski, the director of the Office of Economic and Manpower Analysis at the U.S. Military Academy at West Point. In 1999, after noting that the army had failed to meet its recruiting goals for two years running, he charged his staff with the task of finding ways to entice kids to enlist. One idea that surfaced was video games.

As the AA Web site tells it, in "August of 1999 Wardynski presented his concept for the *America's Army* game" to higher brass and "called for a game that combined single-player adventure and first-person multiplayer action genres into an online virtual Army experience." In January 2000, the Modeling, Virtual Environment and Simulation (MOVES) Institute at the Naval Post Graduate School was chosen to "develop the game under the moniker: Army Game Project" (AGP). As a result, the director of MOVES, Michael Zyda, teamed up with Wardynski, and, that same year, MOVES entered into an agreement with the army to "develop a state-of-the-art video game . . . to educate potential recruits on the army's missions and functions and enhance recruiting opportunities." The MOVES Institute, however, was not alone in creating the game. Along the way, such entertainment and gaming industry stalwarts as Epic Games, NVIDIA, the THX Division of Lucasfilm, Dolby Laboratories, Lucasfilm Skywalker Sound, HomeLAN, and GameSpy Industries also took part.

The game became a huge success for the army and hit the very youth demographic it targets for potential recruits, as well as their younger siblings. AA teaches military training, weapons, and tactics by allowing players to "experience" army life—from the on-screen "rigors" of boot camp to blasting away at enemy troops. It soon became one of the five most popular video games played online, boasting more than 2 million registered users. Since then, a plethora of new versions have been released each year "offering," according to the game's Web site, "new training and/or missions to learn from . . . to provide the *America's Army Community* a careful balance between authentic realism with virtual gaming fun!" As of 2007, the army boasted that "more than 8 million players [had] registered to join the *America's Army* experience," participated "in over 205 million hours of online play" and created more than 1,100 fan Web sites around the world. Furthermore, the game had been downloaded over 40 million times and had been ranked as one of the top ten online games for five years in a row.

When America's Army was released, it was reported that the army had spent approximately $6.3 million on the game's development. But according to a 2005 report by the DoD's inspector general, "four separate army organizations" paid out more than $19 million "to fund research and development of the AGP." The report also found that the MOVES Institute made hundreds of thousands of dollars' worth of improper charges, was incapable of performing the work it pledged to carry out, violated appropriations law, disregarded travel regulations, flouted requirements to safeguard property, gave the appearance of nepotism in its hiring practices, "overcharged the Army Game Project for software licenses that benefited other projects, and misallocated contract labor costs."

While the scandal over MOVES' mismanagement was almost totally ignored by the media, some in the press did raise questions about using the allure of digital violence in a specially crafted game to recruit kids. Christopher Chambers, a graduate of the University of Pennsylvania's Wharton School of Business, former army major and the deputy director of development for AA, responded by alternately admitting and denying that the game was a recruiting tool. In answer to criticisms that its scenarios of blood, violence, and killing were excessive, he insisted, "The game is about achieving objectives with the least loss of life." He noted as well that AA "doesn't reward abhorrent behavior, it rewards teamwork." To highlight the difference, Chambers pointed out that a player who frags (assassinates) his drill sergeant instantly materializes inside a jail cell. Killing non-U.S. personnel, however, is perfectly acceptable as long as it's done the army way.

The marines' Close Combat: First to Fight and the navy-produced America's Army were only the tip of the military's video game iceberg. While these games may be recruiting devices masquerading as toys, there was nothing clandestine about the parties involved in their creation. Much less evident is the military's role in Full Spectrum Warrior (FSW), a combat simulator, unveiled for the Microsoft Xbox system in 2004. FSW allows the gamer to act as an army light infantry squad leader conducting operations in "Tazikhstan," a fictional nation (which apparently sounded so much like the real central Asian land of Tajikistan that the name was later changed to the even more fictional "Zekistan"), against a fictional but stereotypical *evildoer* of the Bush age: Mohammad Jabbour Al-Afad, a former guerrilla leader of mujahideen fighters.

So just how was the military involved? The answer lies in Marina del Rey, California, at the Institute for Creative Technologies (ICT), a center within the University of Southern California (USC) system. There, in 1999, the military's growing obsession with video games moved to a new level when Secretary of the Army Louis Caldera signed a five-year, $45-million contract with USC to create ICT, says the center's Web site, "to build a partnership among the entertainment industry, army and academia with the goal of creating synthetic experiences so compelling that participants react as if they are real."

To accomplish their gaming goals, ICT assembled a team fit for the task, including Executive Director David Wertheimer, formerly the executive vice president of the Paramount Television Group (where he established Paramount Digital Entertainment, the studio's

Internet technology group); Creative Director James Korris (also the executive director of USC's Entertainment Technology Center), a veteran television writer; and Cathy Kominos, formerly the deputy director of research at the Pentagon, where she oversaw the Army Basic Research Program, Simulation, Training and Instrumentation Command, and Army High Performance Computing programs.

In 2003, ICT rolled out Full Spectrum Command, a PC-based combat simulator (modeled after a military role-playing board game), which was developed under the watchful eye of military personnel who teach at the Army Infantry School at Fort Benning. Its purpose was to teach the fundamentals of commanding a light infantry company in urban environments. Using such a game made perfect sense, said Korris, because "35 to 40 percent of incoming military recruits are [already] 'gamers.'" Through the efforts of the developer Pandemic Studios and the game publisher THQ, FSC then spawned the civilian version, Full Spectrum Warrior.

The army's expertise and cash made for a highly acclaimed game that garnered a slew of industry awards, including the most nominations and two wins ("Best Original Game" and "Best Simulation Game") at the prestigious Electronic Entertainment Expo 2003 and a ranking as one of the top ten games of 2004 by *PC Gamer* and *Computer Gaming World*. That same year, ICT was richly rewarded for its video game triumphs when it signed a second five-year deal with the army, which more than doubled its first contract. In fact, the $100-million award was the largest research grant ever received by USC.

ICT is the champion among military gaming centers, but it's only one piece of the army's video game/simulation development puzzle. The same year that ICT was founded, a similar but lesser-known initiative called the University XXI Program, a joint endeavor by the University of Texas, Texas A&M, and the army, was created "to support digitization research at Fort Hood, Texas." By mid-2004, the program had reportedly completed twenty-nine projects. One of its main efforts has been "the Digital Warrior training system," which "incorporates advanced gaming technology features and advanced pedagogical features" to "train military battle captains on the use of the digital army Battle Command Systems in military decision making."

In 2004, capitalizing on the success of AA and interest from other government agencies in using similar technologies, the army also created the America's Army Government Applications Office. Located in Cary, North Carolina, AAGA, now known as Virtual Heroes, consists of "a team of 15 video-game creators, simulation specialists and ex-Army personnel"—many of them hailing "from local video-game companies like Interactive Magic, Timeline, Vertis, SouthPeak Interactive, Vicious Cycle Software and Red Storm Entertainment." The head of the office, Jerry Heneghan, is not only "a West Point graduate, who spent 13 years as an Apache [helicopter] pilot," but also "a producer at video-game developer Red Storm Entertainment, best known for its Tom Clancy–branded military simulations."

FULL SPECTRUM DOMINANCE

Still, it's ICT that attracts the most attention, and with good reason. In addition to creating Full Spectrum Command and Full Spectrum Warrior, ICT is involved in a full spectrum of other military projects including Advanced Leadership Training Simulation, a partnership between ICT and the entertainment giant Paramount Pictures designed for training soldiers in crisis management and leadership skills; Think Like a Commander, a collaboration between the army, the Hollywood filmmaking community, and USC researchers that "support[s] leadership development for U.S. Army soldiers" through software applications; Flatworld, a project that melds "Hollywood set design techniques" with virtual reality technology (and, in 2007, was chosen by the marines for use in their Battle Simulation Center at Camp Pendleton, California, and the new Marine Expeditionary Rifle Integration Facility near Quantico, Virginia); and the Joint Fires and Effects Trainer System project, which "creates an immersive, location-based interactive application for the development of trainee leadership and decision-making skills in joint call-for-fire tasks."

When not creating training aids to improve military lethality, ICT is at work making inroads in Hollywood. For example, its Graphics Lab has "collaborated" with Hollywood film producers and visual effects supervisors on such megamovie blockbusters as *The Matrix* and *Spider-Man 2*. The information transfer flows both ways. In 2005, the writer-director John Milius (*Apocalypse Now, A Clear and Present Danger, Red Dawn*) disclosed that, at ICT, he and other Hollywood insiders engaged in "very, very complex war games . . . for the Pentagon, and we still do that." *Die Hard* and *Commando* screenwriter Steven E. de Souza, directors Joseph Zito (*Delta Force One, Missing in Action*), David Fincher (*Fight Club, Se7en*), and Spike Jonze (*Being John Malkovich, Where the Wild Things Are*); Paul Debevec, who helped to create the bullet-time effects in *The Matrix;* and David Ayer, who cowrote the screenplays for *S.W.A.T.* and *The Fast and the Furious* are just some of his Hollywood compatriots who have lent their talents to the Institute for Creative Technologies. As icing on the cake, ICT even brought in *Star Trek* set designer Herman Zimmerman to create its futuristic workspace.

As part of its mission, ICT has recruited some of Hollywood's best creative minds to dream up futuristic weapons, vehicles, equipment, and uniforms for the army. Production designer Ron Cobb (*Star Wars, Aliens, Total Recall*), for example, lent his creative skills to a program to design the army's supersoldier of the future, the Future Force Warrior (FFW). The FFW is to be unlike any other soldier the army has ever sent into battle, having been "built" from the ground up like other sophisticated weapons systems. The concept relies on constructing an integrated system of weapons, armor, camouflage, and electronics that will monitor a soldier's vital signs and the outside environment. Think of it as another step toward Hollywood's longtime sci-fi dream of a fully realized cyborg soldier—an integrated human/machine combat system that, says the military, will transform a man or woman into a "Formidable Warrior in an Invincible Team." Owing to its Hollywood roots, the FFW is, at least, going to look the part.

In June 2003, General Dynamics won a $100-million contract to complete "preliminary and detailed design" for the Future Force Warrior project. By then, toy maker Hasbro, perhaps best known for its G.I. Joe line of action figures, had also received the specifications of the FFW concept. Why Hasbro? Perhaps because the military recognized that the world of children's toys was the place to go for blue-skies thinking or perhaps because the army reportedly patterned its new quick-loading assault weapons on the design of Hasbro's immensely popular Super-Soaker water gun.

This sort of interconnectedness can get confusing—but we've barely scratched the surface. For instance, the navy's Army Game Project was transformed, in 2002, into the army's free, PC-based recruiting tool America's Army, then reengineered into a training simulator for the Secret Service and the navy. In 2005, it was altered again and combined with video game technology from Laser Shot, Pragmatic Solutions, and Zombie Studios—to create the Future Soldier Trainer (FST), a portable system that "helps to provide life-like training" and also generates data that can be "capture[d]" and analyzed by recruiters "to better understand the factors involved in identifying successful recruits and to stem the tide of attrition."

Set up at recruiting stations as well as at various events geared toward potential future soldiers, FST's realistic weapons are used to lure in kids and also to "test the performance" of recruits in basic rifle marksmanship. This type of technology now appears at venues like air shows in the form of the "Virtual Army Experience." Termed "part video game, part theme park ride and part recruiting tool" by the *St. Petersburg Times,* it allows "civilians a taste of street-level combat in Iraq"—by putting them inside Humvee mock-ups facing a movie screen, where "kids and their parents hunc[h] over faux machine guns, blasting insurgents."

In 2005, America's Army also morphed, thanks to gaming big-gun Ubisoft, into a civilian video game for home gaming systems like the Microsoft Xbox and Sony's PlayStation 2. For about forty dollars, Ubisoft's first AA title, Rise of a Soldier, allowed gamers to learn to be riflemen and snipers from "real active-duty Special

"A youngster tries out the Virtual Army Experience," says Army.mil. *Photo by J. D. Leipold. Courtesy of the U.S. Army.*

Forces operatives [who] consulted with game designers." Meanwhile, in September 2006, America's Army rolled out its twenty-second free, downloadable update of the PC game—America's Army: Special Forces; "Overmatch." In addition to allowing players to use high-powered weapons not available in earlier versions, such as Javelin missiles, Overmatch also marked the beginning of a new form of hyperrealism for the series. The game's "America's Army Real Heroes Program" created eight digital avatars of actual U.S. Army soldiers who survived service in Iraq or Afghanistan. These eight specially chosen "heroes" of America's most recent deleterious occupations won't, however, digitally be sent back for another tour of duty. Instead, they'll be found in "an interactive Virtual Recruiting Center" within the game.

And that isn't the only place the heroes will be showing up. A new line of six-inch-tall "army-authorized" action figures is set to be deployed in "major retail outlets." These twenty-first-century G.I. Joes, produced by toy maker Radioactive Clown, will retail for ten to thirteen dollars and are aimed at AA gamers of enlistment age.

Additionally, in 2007, America's Army was also launched as a cell phone–based video game as well as a freestanding, coin-operated arcade game. The latter effort marked a "unique partnership" between the U.S. Army and coin-op maker GLOBAL VR to "create a new communication channel with young Americans." With input from "U.S. Army Subject Matter Experts and with the full cooperation of units of the U.S. Army," the coin-operated America's Army, GLOBAL VR unabashedly announced, was explicitly "designed to immerse the player in the Army culture."

So to recap: America's Army has morphed from a free PC video game, to a military training simulator, to a recruit testing device, to a commercial home video game, to a line of action figures, to a cell phone game, to a coin-op video game—a series of transformations that typifies the shape-shifting nature of today's military-entertainment complex.

TIME WARP: GAMING GOES BACK TO THE FUTURE

The U.S. military's long campaign to *bring the war home* in the form of video games can be traced back to at least 1929, when Edward Link, in the basement of his father's piano factory, used organ bellows to create the first flight simulator. When amusement parks turned down his machine, Link took his invention to the U.S. armed forces. The navy purchased one in 1931. The army followed suit and, in 1934, began using it as a training device. From that point on, the U.S. military has been hooked on simulation. (It's still hooked on Link's company, Link Simulation & Training—now a division of military-corporate powerhouse L3 Communications, which pulled in $5.1 *billion* from the DoD in 2006.)

During World War II, Link's outfit reportedly churned out some ten thousand "blue box" flight trainers—about one every forty-five minutes. Over half a million pilots used the machines during the war. At the same time, the military was learning that simulation technologies could do more than teach men to fly; they could also teach them to kill. Thus was born the "Waller gunnery trainer," a device created, under a navy contract, by Fred Waller, the former head of special effects for Paramount Pictures (who, in the 1950s, adapted the gunnery trainer technology into the widescreen "Cinerama" movie format). This proto–virtual reality simulator allowed troops to shoot "electronic 'guns,'" which sounded and shook like the real thing, at a huge concave movie screen where simulated targets (films of enemy aircraft in flight) were projected. The military soon found that one hour of simulation could effectively replace three to ten hours of actual, live-fire training. Studies of the effectiveness of gunnery simulators continued after the war and by the end of the Korean War, in 1953, the United States was spending $50 million per year on all types of training devices.

In 1951, Ralph Baer, an engineer working for defense contractor Loral Electronics (today part of Lockheed Martin) on "computer components for Navy RADAR systems," dreamed up the idea of home video games, which he termed "interactive TV-based entertainment." In 1958, at Brookhaven National Laboratory, one of the U.S. Department of Energy's nuclear labs, William Higinbotham created the first proto–video game, Tennis for Two— not unlike the later Pong, in which an on-screen blip was batted back and forth—on one of the lab's oscilloscopes.

In 1962, Steve Russell, a young computer programmer at MIT's Artificial Intelligence (AI) Lab, part of Project MAC (a venture funded by the Defense Advanced Research Projects Agency), created Spacewar, a simple shoot-'em-up featuring spaceships, which ran on a seventeen-square-foot PDP-1 "minicomputer" with a rudimentary monitor. Project MAC was staffed by many game-playing programmers (who called themselves *hackers*) and, according to technology writer Howard Rheingold, "was one of the most important meeting grounds of both the AI prodigies of the 1970s and the software designers of the 1980s." Spacewar quickly spread through computer labs across the United States. In his analysis of "war and video games," *From Sun Tzu to Xbox,* Ed Halter notes that by 1963, Stanford University's Computer Studies Department already felt compelled to ban the playing of the game during business hours.

While military-academic labs were pioneering video gaming, a similar revolution was occurring in the facilities of the military electronics firm Sanders Associates (today part of defense contracting giant BAE Systems, which took in $4.7 *billion* from the Pentagon in 2006). After leaving Loral for Sanders, Ralph Baer wrote his first full report on an interactive home video game system. With support and two assistants provided by Sanders, Baer created the hardware for the first home gaming system in 1969. His invention was eventually licensed to Magnavox as the Odyssey. By 1972, it was being peddled to the public, while larger coin-operated cousins were becoming increasingly common in bars, airports, and shopping center–based video game arcades across America.

At roughly the same time, the army also began taking military simulators to the masses. In 1968, at the height of the Vietnam War, the army set up a weapons simulation display at Chicago's Museum of Science and Industry. There, visitors could electronically "fire" an antitank weapon or test their skills with an M-16. The pièce de résistance, however, was a Bell UH-1D "Huey" helicopter simulator that, wrote the *Los Angeles Times,* gave "visitors a chance to fire an electronic machine gun at simulated Vietnamese homes," causing lights to flash when they hit the target. While protests eventually shut down the army exhibit, the air force continued with plans for a similar exhibit that would allow museum goers to "participate in simulated B-52 missions."

Progress now moved along parallel military and civilian tracks. In 1971, Nolan Bushnell, who had first played Spacewar in a University of Utah computer lab, designed his own version of the game and licensed it to a coin-op manufacturer. The game bombed, but the next year, he founded his own company, Atari, and introduced a simpler game, Pong, which was a huge arcade hit. Soon Atari became the largest builder of coin-op video games. In 1974, there were nearly one hundred thousand coin-operated video games across America when Atari introduced Home Pong—its own at-home gaming console—followed in 1977 by the Atari 2600. While Magnavox's Odyssey used transistors and diodes, allowing for only the most rudimentary of graphics, the Atari 2600 was an eight-bit gaming system that used interchangeable cartridges and represented a quantum leap forward in game play. Atari proceeded to make $5 billion over the next five years with a whole range of video games including the military-themed Combat, Air-Sea Battle, and Battlezone.

As new civilian gaming technology surpassed his Odyssey system, Ralph Baer transitioned right back into military work, reengineering a television-based game into a multifaceted weapons-training simulator. Meanwhile, in 1979, the Department of Defense began investing in the geometry engine, a computer-graphics technology that found its way into later home-gaming systems, like the Nintendo 64.

In the late 1970s and early 1980s, the military began using video games in earnest. In 1977, writes Ed Halter, "programmers at the U.S. Army Armor School modified [a tank simulator called] *Panther Plato* into an obscure prototype training system for tank gunners." Panther Plato reportedly inspired Atari's Battlezone, a minimalist 3D tank simulator, released in 1980, that gave gamers a first-person perspective—looking out from a tank, instead of down on it.

The army immediately saw the possibilities of Battlezone, and, in the year it was released, its Training and Doctrine Command (TRACDOC) approached Atari about creating a military version of the game to be used as a training tool. While the game's designer originally balked at collaborating with the armed forces, the lure of military money and a possible gaming gravy train were apparently too much for Atari executives. At Atari's behest and under the auspices of TRACDOC, the designer modified the game into the souped-up Army Battlezone, which used realistic ballistics, enemy vehicles modeled on Soviet tanks and helicopters of the time, and a realistic controller designed, according to video-game expert Lauren Gonzalez, to mimic "the controls of a Bradley Infantry Fighting Vehicle."

The army then even began to talk of "commission[ing] games to be produced especially for them," while Atari's chief of R&D showed that the technology transfer was a two-way street: the company, he said, was looking into a high-tech helmet like the one "used by military helicopter gunners, which tracks the gunner's eyeballs and aims the weapon where the gunner is looking." (The system Atari created, which actually tracked forehead movement, caused headaches and was never released.)

By 1982, the combined annual take from arcade and home video games was an estimated $7 billion. That same year, the *New York Times* reported that the military was "trying to decide who might best sit at the controls of their evolving electronic weapons," and, not surprisingly, video gamers came to mind since "preliminary tests cited by some experts suggest[ed] that pilots with good psychomotor skills also do well at video games." The *Times* also included a caveat, "No one is considering Pac[-]Man yet as a recruiting device." This wasn't exactly true.

In 1981, an officer from the army's Training Support Center at Fort Eustis, Virginia, proclaimed, "If there's a kid who can score 100,000 points on one of those games right off the start, isn't that the kind of young man who has the hand-eye coordination that could lead to a bright future as a gunner?" The answer was, apparently, a resounding *yes!*

The next year, Brigadier General Winfield Scott Harpe, the commander of the Air Force Recruiting Service, noted that "all those kids buying video games" were his target audience: "The kids who play video games today will be the fliers of tomorrow." That same year, the navy's top recruiter revealed her secret: stalking arcades and plunking down quarters to give potential enlistees free games. "After we played the game," she said, "I'd start asking them what their plans were after high school . . . [t]hen I'd point out that the navy was a highly electronic organization. I knew they'd be into that, because they were into the games." In 1983, the military's commander in chief, Ronald Reagan, speaking at Walt Disney's futuristic EPCOT Center, put the icing on the cake when he likened the military's computerized cockpits to the video game screens mesmerizing American youth. "Watch a 12-year-old take evasive action and score multiple hits while playing Space Invaders," he exclaimed, "and you will appreciate the skills of tomorrow's pilot."

By the early 1980s, military simulators used by actual pilots had become ever more sophisticated and pricey as military-corporate powerhouses like IBM and McDonnell Douglas came to dominate the field. These huge high-tech stand-alone units (like Link's forty-three-ton B-52 bomber simulator) often cost significantly more than the weapons systems they were designed to mimic (such as $35-million simulators for aircraft that cost $18 million) and still only trained personnel in specific tasks, like landing on the deck of an aircraft carrier. The Defense Advanced Research Project Agency (DARPA) called on Captain Jack Thorpe, a flight-training research scientist with the air force, to assess the situation. In 1978, Thorpe posited that simulators should focus on collective training. As such, tens or even hundreds of simulators would need to be networked together—at a time when the air force could barely manage to connect two.

DARPA approved Thorpe's long-term plan to create the SIMulator NETworking, or SIMNET, project using video game and entertainment industry technology. As a result, in 1982, a year in which a Department of Defense catalog listed an inventory of "363 war games, simulations, exercises and models," Thorpe put together a team that included industrial and computer graphics designers to create a network of tank simulators for group training exercises. Elements of the system began their testing phase in 1987, and, soon enough, DARPA's SIMNET enabled the military to war-game, utilizing "approximately 300 players in simulated aircraft and ground combat vehicles located in Europe and the United States on the same virtual battlefield."

Gaming continued elsewhere in the military as well. In 1990, the army's ROTC recruitment campaign began mailing out floppy disks bearing a short video game and interactive quiz to "25,000 college-bound high school seniors" in an effort to stem falling enrollments. That same year, SIMNET became fully operational just in time to provide a virtual training environment for American troops who would soon fight in the first Gulf War. Another computer war-gaming milestone was achieved during the summer of 1990 when the military began *playing* the coming war, digitally, in the form of a simulation known as Internal Look. In his memoir, the former commander of U.S. forces, General H. Norman Schwarzkopf, wrote that at U.S. Central Command (CENTCOM):

> We played Internal Look in late July 1990, setting up a mock headquarters complete with computers and communication gear . . . As the exercise got underway, the movements of Iraq's real-world ground and air forces eerily paralleled the imaginary scenario of the game . . . As the game began, the message center also passed along routine intelligence bulletins about the *real* Middle East. Those concerning Iraq were so similar to the game dispatches that the message center ended up having to stamp the fictional reports with a prominent disclaimer: "Exercise Only."

But Internal Look did not go off without a hitch. The day of Saddam Hussein's invasion of Kuwait (which he allegedly war-gamed using an American-made simulator), Michael Macedonia— the son of the man who was reportedly one of the first to "introduce computer war games to the Pentagon" and who, himself, had been "experimenting with computerized war games" since the 1980s and had become a manager of information systems for one of the military's electronic warfare centers—was flown to CENTCOM to work on its crashing computer system. Eventually, the whole CENTCOM gaming center was packed off to Saudi Arabia, where, during the war, plans developed as part of the war game were utilized in actual military operations.

During the conflict, American audiences were captivated by images of technowar made bloodless in the green glow of night-vision photography and carefully selected (and totally unrepresentative) video images, provided by the Pentagon to breathless television networks, of "smart" weaponry hitting targets. Given this atmosphere, it is hardly surprising that Douglas Kellner, author of *The Persian Gulf TV War,* termed the conflict a "high-tech cyberspectacle" that literally took "TV viewers into a new cyberspace, a realm of experience with which many viewers were familiar through video and computer games . . . The fact that bombs were falling on Iraqi civilians and destroying their homes and social infrastructure was also obscured by the Nintendo-like video images of the pyrotechnics of modern warfare."

At roughly the same time as U.S. pilots in F-15 fighters were slaughtering Iraqis from the skies, Americans on the home front were doing the same thing—sort of. After selling more than one million copies to PC gamers, MicroProse Software began offering its F-15 Strike Eagle as a coin-operated arcade game. "For an initial investment of 50 cents, you can fly the plane that bombed Baghdad" was the way one newspaper article described the game. (One of the F-15 Strike Eagle arcade games was even flown to Saudi Arabia for U.S. personnel to play during downtime.) Not only was F-15 Strike Eagle reported to be a civilian "version of an actual combat flight simulator," but MicroProse's president, John W. "Wild Bill" Stealey, was an Air Force Academy graduate and a retired USAF lieutenant colonel whose previous job had been "advising the Joint Chiefs of Staff."

Just days into the ground combat portion of the Gulf War, the Battle of 73 Easting pitted American armored vehicles against a much larger Iraqi tank force. The U.S. troops, who had trained using the SIMNET system, routed the Iraqis. Within days, the military began turning the actual battle into a digital simulation for use with SIMNET. Intensive debriefing sessions with 150 veterans of the battle were undertaken. Then DARPA personnel went out onto the battlefield with the veterans, surveying tank tracks and burned-out Iraqi vehicles, as the veterans walked them through each individual segment of the clash. Additionally, radio communications, satellite photos, and "black boxes" from U.S. tanks were used to gather even more details. Nine months after the actual combat took place, a digital recreation of the Battle of 73 Easting was premiered for high-ranking military personnel. Here was the culmination of Thorpe's efforts to create a networked system that would allow troops to train for future wars using the new technology combined with accurate historical data.

In late 1993, with the green glow of Gulf War victory already fading, id Software introduced the video game Doom. Gamers soon

began modifying shareware copies of this ultraviolent, ultrapopu-
lar first-person shooter, prompting id to release editing software
the next year. The ability to customize Doom caught the attention
of members of the Marine Corps Modeling and Simulation Management Office who had been tasked by the corps' Commandant
Charles Krulak with utilizing "personal computer (PC)-based war
games" to help the marines "develop decision making skills, particularly when live training time and opportunities are limited."

Acting on Krulak's directive, the marines' modeling crew nixed
Doom's fantasy weapons and labyrinthine locale and, in three
months' time, developed Marine Doom, a game that included
only actual Marine Corps weaponry and realistic environments.
Krulak liked what he saw and, in 1997, approved the game. "It's
not meant to replace field time. It never will," said project officer
Lieutenant Scott Barnett. "But there's a whole lot more that you
can do with this tool. The fun factor is very important. That's what
makes our Marines want to use it. But it's an honest-to-God training tool. You can do mission rehearsal, mission planning."

Doom was just one game in the corps' *arsenal*. The marines were
also playing such games as Harpoon2, Tigers on the Prowl, Operation Crusader, Patriot, and another id first-person shooter, Quake,
as well as hosting Doom tournaments. They also signed up with
Good Times Interactive for a follow-up game: Battlesight Zero—
exchanging their "input and combat expertise" for a $1-million
investment in the game. But the marines were out for even bigger game.

Back in 1990, two members of the original SIMNET project
team, Warren Katz and John Morrison, had founded MÄK Technologies to develop "software to link, simulate and visualize the
virtual world." In 1997, the Marine Corps signed a deal with MÄK.
No longer content to dabble only in off-the-shelf software, the
marines tasked MÄK to create the "first video game to be co-
funded and co-developed by the Department of Defense . . . and
the entertainment industry." Dubbed MEU-31 (an elite Marine
Expeditionary Unit), the game, according to Katz, represented "a
major step for the DOD in that they are recognizing the benefits of

collaborating with a commercial video game publisher from the beginning of the game design process. This will result in a video game which is much more realistic than any other game ever produced for this genre, making its commercial success highly likely, while at the same time, giving the DOD the cost benefit of unusually large volume sales for a military training device."

The year that it inked the deal with the marines, MÄK (whose other customers include such defense industry stalwarts as ITT Industries, Boeing, Lockheed Martin, and Raytheon) joined forces with Zombie Studios—a company headed by Mark Long (a retired U.S. army officer who had worked at General Dynamics' Combined Arms Systems Engineering Laboratory) and Joanna Alexander (who had served as deputy chair of a Pentagon committee advising military officials on issues relating to technology transfer and virtual reality)—to publish the M1A2 Abrams battle tank simulator, Spearhead. Soon after, the army contracted with MÄK to create Spearhead II, a SIMNET simulator designed to train tank crews and commanders in tactical decision making. Meanwhile, the navy began using 688(I) Hunter/Killer: Submarine Simulation Game. Speaking about such collaborations between the gaming industry and military, Katz accurately predicted that "we are going to see more and more of this during the next several years. This is going to become a big trend."

Soon enough, the Army Game Project was at work on America's Army, ICT was developing its Full Spectrum Command simulator and, in 2001, the DoD modified Tom Clancy's Rainbow Six: Rogue Spear, by game maker Ubisoft, to train military personnel on how to conduct small-unit military operations in urban terrain. The rest followed, leading to the current continuous military gaming/simulation loop where commercial video games are adopted as military training aids and military simulators are reengineered into civilian gaming moneymakers in all sorts of strange and confusing ways.

For instance, in 2005, as video games climbed to a $25-billion-a-year global industry, a Web site created, developed, and designed with grants from both the DoD's Joint Advanced Distributed Learning Co-Laboratory and DARPA reported that the U.S. military was using at least fifty-five video games and simulators. The list spanned the gamut from custom-made training programs such as Anti-Terrorism Force Protection by Will Interactive; to military-produced fare like America's Army and commercial off-the-shelf games like the Microsoft Flight Simulator (used by the navy), Breakaway Games' Peloponnesian War (a strategy game used at the National Defense University), Activision's Soldier of Fortune, and Electronic Arts' Medal of Honor (both used by the marines at their Infantry Cognitive Skills Labs) as well as Konami's Air Force Delta Storm and Blizzard's Starcraft, where players take part in a "galactic conflict between three different species on the edge of known space" (both used by the air force).

On the civilian side, that very same year, you could play the army's navy-designed America's Army on a personal computer (PC) for free, or play a pay-to-play version on the Microsoft Xbox system, or even use your PC to take part in the Special Teams Challenge at the GoArmy.com Web site. Similarly, you could use your home computer to download the National Guard's Guard Force, "a real-time strategy game using modern military equipment and units . . . [operating] in snow covered mountains and lush jungles, performing covert assaults, counter-insurgency and rescue." You could run the navy's Navy Training Exercise Strike and Retrieve game on your PC, or occupy yourself with SOCOM II: U.S. Navy SEALs, a game produced with the assistance of the Naval Special Warfare Command, or SOCOM 3, for which the navy provided "technical support," on your Sony PlayStation 2 system. At air force recruiting centers you could "fly" the USAF's F-22 Raptor, a Predator unmanned drone, or a C-17 Globemaster III transport while playing USAF: Air Dominance. At the Air National Guard Web site there were three possibilities: Sky Tank, Air Battle, or Parachute Mission. And if you weren't already sated, the good news was: The Marine Corps simulator-turned-video game Close Combat: First to Fight had just released for play on the Xbox and PCs.

CHILDREN AT WORK, DO NOT DISTURB

The U.S. intelligence agencies have not been shy about forging their own connections with the toy, entertainment, and video game industries. In recent years, ICT has hooked up with the CIA to develop a game to help agency "analysts think like terrorists." CIA spokesman Mark Mansfield explained, "For out-of-the-box thinking, we are reaching out to academics, think tanks and external research institutes that are critical in the fight against terrorism."

CIA counterterrorism officials who traveled to ICT headquarters were given VIP tours of Hollywood movie studios. In 2003, the agency landed Jennifer Garner, then playing the role of a secret agent on ABC's popular CIA drama *Alias,* to star in its recruitment videos. Said the CIA's liaison to the entertainment industry, Chase Brandon, "If Jennifer ever decides she doesn't want to wear dark glasses of the celebrity status, she can put on dark glasses and be a spy. She's got what it takes." In the meantime, Michael Clarke Duncan, who costarred with Garner in the movie *Daredevil,* lent his voice to the navy-aided Sony PlayStation game, SOCOM II: U.S. Navy SEALs. That same year *Alias* itself, complete with Garner's voice, was turned into a video game by Acclaim Entertainment, a company that had donated "10,000 hand-held video games" to U.S. troops sent to the Middle East during the Gulf War.

Another game to emerge alongside SOCOM II and Alias was Kuma War, developed by Kuma Reality Games in cooperation with the U.S. military. Billed as the first commercial shooter game that allowed players to re-create actual military missions—with each combat operation introduced by television footage and a cable-style news anchor—the game also provided players with "video news shows" and "extensive intelligence gathered from news sources around the world." Like any good entity in the Complex, Kuma linked itself to the military through the Pentagon's revolving door of employment: A retired Marine Corps major general serves as one of its top consultants. In fact, Kuma boasts a board of military veteran advisers "whose job it is to make sure the missions

[they] put out are as realistic as possible." By mid-2007, Kuma had released eighty-three "missions," ranging from the killing of Saddam Hussein's sons, "Uday and Qusay's Last Stand" (mission 1); to a digital re-creation of the May 22, 2004, "Battle in Sadr City" between U.S. troops and the Iraqi cleric Muqtada al-Sadr's Mahdi army (mission 16) and, owing to U.S. failures in Iraq, "Sadr City (Revisited)" (mission 78).

If its Web site is to be believed, Kuma War has a hard-core following not only among civilian gamers but also among soldiers. One Iraq War veteran quoted on Kuma's home page confesses, "This game actually makes me flash back and think about the war . . . But that's not necessarily bad. Being that I will be going back to Iraq for a 3RD tour, I'll say that it's much better fighting from my PC behind a desk then [sic] actually slinging lead at each other."

Of course, such games are increasingly seen as a means to prepare soldiers to *sling lead.* In 2006, David Bartlett, the former head of the Defense Modeling and Simulation Office and a cocreator of Marine Doom, told the *Washington Post,* "The technology in games has facilitated a revolution in the art of warfare." One Iraq War veteran cited in the article suggested that video gaming was key to his ability to unleash a .50-caliber machine gun on a "human enemy." A devotee of the video games Full Spectrum Warrior and Halo 2, he proclaimed, "It felt like I was in a big video game. It didn't even faze me, shooting back. It was just natural instinct. *Boom! Boom! Boom! Boom!* . . . I couldn't believe I was seeing this. It was like 'Halo.' It didn't even seem real, but it was real."

"When the time came for him to fire his weapon," Bartlett commented, "he was ready to do that. And capable of doing that. His experience leading up to that time, through on-the-ground training and playing 'Halo' and whatever else, enabled him to execute. His situation awareness was up. He knew what he had to do. He had done it before—or something like it up to that point." Lieutenant Colonel Scott Sutton, the director of the technology division at Quantico Marine Base, agreed, asserting that troops raised on first-person shooters "probably feel less inhibited, down in their primal level, pointing their weapons at somebody," and noted that video game pretraining "provides a better foundation for us to work with."

The relationship between video games and the military, however, hardly ends there. Gaming *machinery* is allowing troops to operate other military "toys" with minimal training. For instance, the Dragon Runner, a small remote-controlled carlike vehicle designed to travel inside buildings and spy for marine patrols waiting outside, is a case in point. Developed by researchers from the Naval Research Laboratory and Carnegie Mellon University's Robotics Institute, working with the Marine Corps' Warfighting Laboratory, the Dragon Runner is guided by a six-button keypad, modeled after Sony's PlayStation 2 (PS2) video game controller. Major Greg Heines, a marine attached to the Warfighting Lab project, says it was chosen because "that's what these 18-, 19-year-old Marines have been playing with pretty much all of their lives, [so they] will pick up [how to drive the Dragon Runner] in a few minutes." Another marine had a similar experience with a prototype Gladiator Tactical Unmanned Ground Vehicle—a four-foot-tall, 1,600-pound "tele-operated, semi-autonomous vehicle"—noting that it took him just five minutes to learn to operate the vehicle. "It's like a video game. You have a joystick and you drive it forward, backward, right or left."

According to Michael Macedonia, when soldiers were asked what they'd like to use to control guided missiles, they picked the PS2 controller. As a result, it has, indeed, become the preferred model device. The army's iRobot PackBot EOD Tactical Mobile Robot is also guided by what "looks like the hand-held video game control often attached to a PlayStation." One soldier using the robot proudly told the Associated Press, "My family thought it was a waste of time playing those video games. I finally proved them wrong."

Colonel John Burke, the project manager for the army's Unmanned Aerial Vehicle (UAV) Program, noted that teenage troops are able to learn to launch and fly the army's drones in a mere eight hours time because the controls look "very much like a PlayStation controller." In 2005, *Wired* magazine's military tech expert Noah Shachtman described a nineteen-year-old army UAV operator, noting that he had "been prepping for the job since he was a kid: He plays video games. A lot of video games. Back in the barracks he spends downtime with an Xbox and a PlayStation. When he first slid behind the controls of a Shadow UAV, the point and click operation turned out to work much the same way." The air force even recruits by using a war-as-play/weapons-as-toys approach, with a Web site feature showing a young camo-clad airman standing in front of a UAV, beneath graphics that equate flying drones to playing with radio-controlled toy airplanes.

The military has also reworked one of its most successful civilian toys for training purposes. America's Army technology has now been reengineered to train soldiers (perhaps the very same people recruited by the game in the first place) in the use of remotecontrolled Talon robots for operations in Iraq and Afghanistan. Additionally, Virtual Heroes and Zombie Studios collaborated to reengineer AA to teach soldiers "convoy skills" for missions like those in Iraq, including "VIP transport, pick-ups or drop-offs, and trying to reach a checkpoint to hand off their vehicles."

On the civilian front, beginning in 2007, America's Army also became, not surprisingly, one of the games featured at the Army Gaming Championships—a tournament for hard-core gamers, ages seventeen and up, where over $200,000 in cash and prizes were up for grabs. There was, however, a catch. To take home the loot for playing any of the twelve martial games such as AA, Halo 2, Call to Duty 3, Tom Clancy's Rainbow Six: Vegas, or the ultraviolent Gears of War, players had to give permission to be contacted by an army recruiter. That same year, the army signed a $2-million deal with Global Gaming League—an online gaming community— as a way to tap into the 9.2 million players (80 percent of whom are seventeen- to twenty-four-year-old males) who visit the site each month.

Today, the military, toy, and gaming worlds are completely entangled, and the future promises only more interpenetrations and complex collaborations that would have made Dwight Eisenhower's head spin. With defense spending hovering in the $555-billion range annually (and actual national security expenditures edging toward the $1-trillion mark); with the U.S. toy industry registering $22 billion annually, annual video game software and hardware sales in the U.S. topping $12.5 billion, and U.S. sales of PC games reaching over $1 billion each year; with an overstretched all-volunteer military, multiple unpopular wars abroad, a Global War on Terror in full swing, and no outcry from the public over the militarization of popular culture—who knows what the future holds?

Certainly, the day is not far off when most potential U.S. troops will have grown up playing commercial video games that were created by the military as training simulators; will be recruited, at least in part, through video games; will be tested, postenlistment, on advanced video game systems; will be trained using simulators, which will later be turned into video games, or on reconfigured versions of the very same games used to recruit them or that they played as kids; will be taught to pilot vehicles using devices resembling commercial video game controllers; and then, after a long day of real-life *war-gaming* head back to their quarters to kick back and play the latest PlayStation or Xbox games created with or sponsored by their own, or another, branch of the armed forces.

More and more toys are now poised to become clandestine combat teaching tools, and more and more simulators are destined to be tomorrow's toys. And what of America's children and young adults in all this? How will they be affected by the dazzling set of military training devices now landing in their living rooms and on their PCs, produced by video game giants under the watchful eyes of the Pentagon? After all, what these games offer is less a matter of simple military indoctrination and more like a near immersion in a virtual world of war, where armed conflict is not the last, but the first—and indeed the only—resort.

13

PIMP MY PENTAGON

At least as far back as the 1970s, heading to the hot-rod track meant a Pentagon sales pitch. There, one of the top drag racers, Don "Snake" Prudhomme, smoked the competition in a nitro-fueled, red-white-and-blue funny car with *U.S. Army* painted on its sides. When not burning up the track, Prudhomme visited high schools on behalf of the army, signing autographs for the kids ogling his recruiting poster on wheels. So as not to be left in the dust, the Marine Corps got into the act, paying out handsome sums to get its recruiting slogans painted on cars racing in International Hot Rod Association competitions, while the navy cheaped out and took only donated advertising space.

Back then, the idea was to grab the attention of young gearheads and get them to join the armed forces. Today the Pentagon has really revved up its engine. Head down to the drag strip today, and you'll find the screaming yellow and black "U.S. Army Top Fuel Dragster" driven by Tony "The Sarge" Schumacher, the 2004, 2005, 2006, and 2007 National Hot Rod Association (NHRA) Top Fuel Champion.

An "honorary sergeant" in the army, according to *Government Executive* magazine, Schumacher, who has been on the Pentagon payroll since 2000, is unabashed about his role. He told the Webzine *Gelf*, in no uncertain terms, "I'm a recruiter." And each week, uniformed army recruiters give cash-strapped youths free tickets to racetracks to see the army dragster and hear a pitch from their hot-rod idol. There, Schumacher fills their heads with talk of good-paying jobs, apparently successfully selling the army, even in a time of disastrous war experiences.

Schumacher isn't alone. The army racing team also includes Angelle Sampey and Antron Brown, two NHRA Pro Stock Bike "pilots"—motorcycle drag racers. These easy-riding recruiters tool down the track on army-emblazoned Suzuki bikes to similar effect.

The best-known component of army racing is its Chevy Monte Carlo driven in the NASCAR NEXTEL Cup series. NASCAR has long been in league with the U.S. military—the close relationship between the two having been formed by the association's founder, Bill France. But only since 2000 has the Pentagon been putting on the hard sell to racing fans. In addition to the army, the other services—the navy, marines, and air force—all have NASCAR teams. There are even Army National Guard/Army Reserve, Marine Corps Reserve, and Air National Guard/Air Force Reserve cars. In 2004, the Pentagon's lavish ad dollars, aimed at 8.5 million NASCAR fans between the ages of eighteen and twenty-four, generated a reported "30,000 qualified leads" for the army alone. In 2005, the military spent more than $38 million in taxpayer money to fund various racecars.

Along with eyeing up young red-state whites with NASCAR dreams, the Pentagon has also turned to pimped-out rides to secure more minority cannon fodder.

In an effort to nab "urban" recruits, the army hits the streets with a totally pimped-out ride. Bearing the army's Spanish-language advertising slogan, the "Yo Soy El Army" ("I Am the Army") custom Hummer H2 is billed as "one tricked out vehicle." While the war in Iraq showed that the army's Humvees were no match for Iraqi guerrillas' roadside bombs, stateside the army tools around in a chrome-laden, custom-painted Hummer H2 with a custom leather interior and a souped-up entertainment system—complete with a Pioneer in-dash CD/DVD player and a state-of-the-art video package boasting a total of fifteen TV screens.

The "Yo Soy El Army" Hummers, and the similarly painted, flashy big rigs that carry them, are actually part of a multipronged

The army Chevrolet and navy Dodge at the 2005 USG Sheetrock 400 NASCAR race. *Photo Courtesy of the U.S. Army.*

recruiting offensive undertaken for the army by Latino Sports & Entertainment Marketing of San Diego. This firm boasts of being "the Latinos speaking to Latinos company" in the field of "Latino sports and entertainment marketing, event management, experiential marketing and consulting."

Tasked with generating "qualified leads of potential Latino recruits nationwide," in just one year's time, Latino Sports & Entertainment Marketing arranged for the army to sponsor fairs on sixteen college campuses to tout active duty and ROTC service and set up booths at forty Major League Soccer games. (The army deploys at plenty of other "Latino" events as well. For instance, 2007 marked the third straight year that it had sponsored a day of festivities, and set up a recruiting station, at the Tropical Music Festival, a Latin music concert series in the Bronx.) The Hummers, however, were its pièce de résistance. The two customized vehicles, according to the company, "toured markets across the country and were showcased at car shows, Hispanic festivals, concerts and sporting events" driving "traffic to the Armys [*sic*] information booth."

All told, Latino Sports & Entertainment Marketing claimed that its staff "traveled more than one hundred thousand miles covering more that twenty-five key cities across the nation, providing the army exposure at hundreds of events," reducing the army's cost-per-lead for recruiters by 29 percent and increasing lead generation by 31 percent. One high school student, in an excerpt from an editorial for his school newspaper, republished in the *Los Angeles Times*, recounted his underwhelming real-world experience with the effort: "The army managed to get a Hummer rolling on 24-inch dubs, blasting rap, lined with flames on the side, outside of Room C161." As for the pitch? "Dressed in army uniforms, recruiters stood outside telling people that if they signed up, they [would] receive a T-shirt that said, in Spanish, 'YO SOY EL ARMY.'"

In addition to these Hummers, the army deploys an entire fleet of flashy SUVs to college campuses and various youth-oriented special events.

Meanwhile, the marines roll out their own flashy Hummer, which they call an "Enhanced Marketing Vehicle," at events like college football's Outback Bowl. They also take it to other high-traffic venues, where, parked alongside an "inflatable giant-sized drill instructor," its custom paint-job, "fully loaded sound system, and a video game console" are designed to draw youths to four marine drill instructors manning what they call an "enhanced area canvassing event."

The air force, too, has taken a similar tack. In late 2002, it got its own pimped-out ride: a 2003 GMC Yukon XL that it calls the "RAPTOR SUV"—after the air force's F-22 fighter jet. Custom-painted in blue, white, and gray, replete with air force logos, backlit grills, custom rims, leather interiors, entertainment centers featuring a forty-two-inch plasma-screen TV, DVD player, full-range sound system and even a Sony PS2 thrown in for good measure, thirty-one RAPTOR SUVs (along with tricked-out trailers "carrying 6 flight simulators each" and stand-alone miniature models of the F-22) function as mobile marketing tools. Deployed across the country, they are meant to attract youth at local fairs, air shows, and extreme sports events like the Dew Action Sports Tour. In 2005, they were even joined by a tricked-out motorcycle—courtesy of Orange County Choppers—to be "used at recruiting venues." The ten-foot-long bike, modeled after the F-22 Raptor, sports a thirty-one-cubic-inch displacement engine capable of generating 150 horsepower, custom rims bearing the air force logo, rearview mirrors in the shape of jets, and even an attached cannon round.

When the military sends its tricked-out rides to events like Hot Import Nights—a popular traveling car show and entertainment expo—it tends to send along eye candy, as well. At the 2003 Winter Music Conference in Miami Beach, the army deployed, according to Celeste Delgado and David Holthouse of the *Miami New Times,* "a street team of hot enlisted chicks in camouflage who distributed mock dog tags to spring breakers." These women are generally "enlisted" in modeling agencies and hired out to the army. For example, one army model (whose turn-ons include "chivalry" and "exotic cars" and turn-offs include "ugly/slow cars" and "accents that are not British/French/Italian"), said her agency arranged for her to join six to eight other women, all wearing black T-shirts emblazoned with the army logo, at an interactive "setup designed to raise awareness and possibly recruit people." Her job "was to attract people to the setup and guide people through the display . . . instructing people how to operate an interactive video game, signing people up to receive information, and operating a machine to make faux dog tags for people with their name on it."

Hundreds of thousands of young people attend these "lifestyle" car shows to ogle fast cars and sexy models. Between June 2006 and March 2007, the army appeared at thirteen Hot Import Nights events, while its National Guard brethren sponsored all twenty Hot Import Nights events in 2006 and signed on for twenty-one more in 2007.

And Hot Import Nights is only one of the car shows that the military now frequents. In 2004, the "Yo Soy El Army" Hummer also showed up at the Las Vegas Lowrider Supershow—a sign of a new military emphasis on the once California-centric phenomenon of shined-up, tricked-out customized cars. With the success of MTV's *Pimp My Ride* (a TV show hosted by rap artist Xzibit) and *Lowrider* magazine's eponymous Sony PlayStation 2 video game, the military decided to make more plays for the attention of lowrider aficionados and wannabes, sponsoring *Lowrider* magazine's fifteen-city Evolution Tour.

The writer Julianne Shepherd notes that by enmeshing itself in a traditionally Latino car subculture, originally pioneered by Mexican American migrant workers, "the US army is using *Lowrider* and lowrider culture to specifically target and recruit Hispanic Americans." The army has also specifically targeted African American youths. As part of its "Takin' It to the Streets" effort, which was created by an African American–owned PR firm, the army sends its Hummer-driving street teams to such events as NAACP conventions, "select schools, malls and national events" like Missouri's Black Expo and Spring Bling, Black Entertainment Television's signature spring break hip-hop festival. There, a motion simulator display, a rock wall, and a customized "Army of One" Hummer, specially outfitted with a regulation-height basketball hoop attached to the rear of the vehicle, are used to catch the attention, says the Pentagon, of "African-American High School, College and workforce prospects."

In 2007, the army cosponsored the Essence Music Festival—an event celebrating African American music and culture—and also sent its Elite Step Team. And, as *Crain's New York Business* noted that same year, the army even began stocking its Hummers with "rap CDs and other goodies," while its street teams hit the pavement with "glossy fliers as if they were promoting a rap concert."

Obviously, the Pentagon's automotive-themed efforts are effective in getting at least some hot-rod-loving, NASCAR-watching or lowrider-obsessed youths to consider the military. Of course, as soldiers, they end up riding in less flashy, nontricked-out vehicles— unless you count the scavenged scrap-metal armor plating that soldiers were forced to weld onto unarmored Humvees in Iraq.

PART V

THE COMPLEX
GOES RECRUITING

In June 2007, the army quietly pulled one of its television recruitment ads after receiving a complaint that it contained a blatant lie about the type of real-world training available in the army. The commercial claimed that army training could set soldiers up to become civilian pharmacists when their tours ended. Not so, according to David Work, the former president of the National Association of Boards of Pharmacy, who noted that the army doesn't offer a six-year program at a school of pharmacy necessary to become accredited. "They knowingly, intentionally put a lie out there, only to get a teenager to sign up," said Work. "Any teenager will find a six-figure job attractive."

Work was right. Slick productions, promises of big money, and lies have become the cornerstones of military recruitment efforts. For the Complex to function and wage wars, it needs bodies, and it's increasingly pulling out the stops to get them—from forays into the world of online social networking and the creation of civilian-looking career guidance Web sites to recruiters who sign up autistic teens, recent psychiatric ward discharges, and convicted felons. These days, the only real question is: How low will they go?

14

AN ARMY OF (NO) ONE

With the U.S. military bogged down in a seemingly interminable mission in Iraq, supposedly "accomplished" years earlier, the armed forces are in tough recruiting shape, but you wouldn't know it from the Internet. There it's still a be-all-that-you-can-be world of advanced career choices, peaceful pursuits, and risk-free excitement— a digital world increasingly dotted with Web sites of every sort, lying in wait for curious teens (or their exasperated parents) who might be surfing by.

In addition to raising the maximum enlistment age (to 42), no longer dismissing new recruits out of hand for "drug abuse, alcohol, poor fitness and pregnancy," allowing those with criminal records in, and wielding hefty sign-up bonuses along with tens of thousands of dollars of "mortgage assistance" to lure the cash-strapped into enlisting in the *all-volunteer* military, a favorite method of bolstering the rolls is the time-tested targeting of teens to fill the ranks.

To this end, the Pentagon engages companies like Teenage Research Unlimited. TRV, which received more than $720,000 from the Pentagon from 2001 to 2006, offers "clients virtually unlimited methods for researching teens," in order to get inside kids' heads. In addition, the Department of Defense—through its Joint Advertising Market Research and Studies (JAMRS) program and Mullen Advertising, as well as Mullen's subcontractor, the private marketing firm BeNow, has created what is "arguably the largest repository of 16–25 year-old youth data in the country, containing roughly 30 million records." One key component of this data set is JAMRS's "High School Master File," which contains a list, updated five times a year, of the "contact information [for] . . . about 90 percent of the high school population." Armed with "names, birth dates, addresses, Social Security numbers, individuals' e-mail addresses, ethnicity, telephone numbers, students' grade-point averages, field of academic study and other data," the Pentagon now has excellent ways of accurately targeting teens.

The Pentagon has another prime way of gathering information as it seeks out the youth market: military Web sites. In recent years, these have proliferated, functioning not only to attract new, youthful recruits but as interactive multimedia environments where the military can gain *intel* on those potential recruits. Says the fine print on one of the navy's sites: "Information the United States navy collects on you from navy Chat, a Message Board session or e-mail message will be used for the purpose of recruiting enlisted personnel and Officers into the United States Navy and Naval Reserve."

"WE'VE BEEN WAITING FOR YOU"

To win over the kids, multimedia sites are the order of the day. The air force's Airforce.com, one of a number of official USAF sites, uses a banner that announces, "We've been waiting for you," above interactive, eye-catching content designed to appeal to kids, including an inside-the-cockpit shot that shows an air force pilot in a full G-suit looking over his shoulder at another sleek fighter jet. "I used to play tag at the park, now I play tag at Mach 2," reads the caption.

Kids who search the phrase "cool air force" using (2006 DoD contractor) Google's search engine will find another teen-friendly site: U.S. Air Force's DoSomethingAmazing.com—a site devoted to the culture of cool. Viewers are immediately hit by a barrage of eight video clips playing simultaneously—an air force security force attack dog knocking a man to the ground, the massive explosion of a bomb, an F-22 Raptor, men jumping out of planes, and eye-popping aerial acrobatics, among others—offering "never-before-seen" glimpses of what they describe as "awesome skill and fighting power." There's even a link to a "Cool Stuff" section of another air force Web site.

Anyone who follows that "Cool Stuff" link is greeted with a site full of USAF screensaver and desktop wallpaper downloads. There teens can read about Special Ops units; learn about high-tech weaponry; and watch the latest air force TV commercials—such as a montage of extreme snowboarding and supersonic bombers set to a driving rock beat. The Web site even offers a direct link to "get in touch," by e-mail, not with a recruiter, mind you, but with "an adviser." And those who can't wait can even log in for an instant live chat with a recrui . . . um, adviser.

IN THE NAVY . . .

In the war for America's youth, the navy does the air force one better with an interactive site that allows teens access to a "Virtual Recruiter." Its "About You" interactive feature has a "Who Am I?" section where you view images and pick the one that best represents you. Because I chose, among others, a picture of people playing volleyball over one of someone working out alone on a treadmill and clicked on a picture of a man in a jacket and tie getting sprayed by a sprinkler over a guy buying movie tickets, the navy declared: "You're down with the team. You'll do whatever it takes . . . You have a brain and you use it. You read books without pictures in them . . . You like living on the edge. Snowboarding, rollerblading and mountain biking . . . you're down with extreme sports . . . You're into living life and doing your thing."

Then I moved on to the navy's trademarked "Life Accelerator v2.0," which "maps your likes and dislikes to six skill and interest profiles." After I told it that, among other things, I'd like to write a short story, dislike "restoring antique furniture," and would enjoy a career as a journalist, the navy put together a bar chart for me—displaying my supposed aptitudes—that I was encouraged to "share with [my] family and friends" or "send to a recruiter." Finally, the navy's interactive recruiter gave me a chance to decide who I could be after my first "12 to 18 months in the navy" or even what I could "accomplish after 4 to 6 years."

In the near term, the navy said I'd be heading off for "Warfare Designation Training" and "might be jumping off a 12-foot tower in full gear" or even receiving small-arms training. And while I could be "learning about and training onboard a nuclear attack submarine or flying an F/A-18 Super Hornet," such activities paled in comparison to the fun I'd have during "time off on weekends and at day's end (liberty)," including: intramural sports, hitching rides on military aircraft, scuba diving, and deep-sea fishing. The navy even offered me a downloadable video game, Training Exercise: Strike & Retrieve, reinforcing the fun, fun, fun ethos.

The navy also offers a bit of fun for the more junior set. Ostensibly created to provide information about oceanography, Neptune's Web is a cartoon-laden site geared toward the elementary and middle school crowd. However, the benign-sounding Naval Meteorology and Oceanography Command that created the site isn't dedicated to scientific discovery but instead tasked with providing "an asymmetric war fighting advantage through the application of Oceanographic sciences." Not surprisingly, the site makes a none-too-subtle appeal to youngsters: "Today's high-tech Navy is focused . . . on the advanced technology of tomorrow . . . a tomorrow that can include you."

THE MARINE CORPS(E)

While the navy's recruiting Web site offers a vision of military life as easy living—from fishing to surfing—the Marine Corps take a different tack. As with all the armed forces Web sites, there's evidence that military marketing research indicates extreme sports play well with likely recruits—so there are the obligatory shots of rock climbing—but above all the marines use their Web site to sell a swagger that's perfect for the volatile male teen itching to kick ass and take names. For this testosterone-drenched segment of the recruitable population, the marines' Web site (amid pictures of rippling muscles and heavy weapons) is locked and loaded with macho hyperbole.

> The Marines are the battering ram that smashes the enemy's door.
>
> Marines are collectively and individually lethal.
>
> We are warriors . . . built to conquer . . . We are fierce in a way no others can be.
>
> You will pose a true threat to your enemies.

In fact, even the marines' digital camouflage uniform is so bad-ass that it alone can "inspire fear in the hearts and minds of all enemies." Still, if the lure of being an arrogant killer in intimidating threads isn't enough, the Web site has one last secret weapon: trinkets. Anyone who requests information for the first time gets a free "Marine Corps carabiner [*sic*] keychain."

AN ARMY OF FUN

While the marines foster a murderous mystique, the navy offers visions of fun in the sun and the air force makes it seem like it's forever playtime, the ground-pounding army's Web world really has it all.

The army offers a plethora of Web sites and Web pages geared toward the youth demographic. There's HOOAH 4 Kids, a Web site filled with games, including a memory card game where the reverse side is, of course, camouflage and an online coloring book that allows for printable patriotic play.

Young science-minded teens searching the Web may stumble across eCYBERMISSION, a site devoted to "a web-based science, math and technology competition for 6th, 7th, 8th and 9th grade teams" that offers the chance to "compete for regional and national awards" and win thousands in savings bonds. You wouldn't know it from the site's name (and it lacks the typical ".mil" Web address), but if you click on a tiny "sponsors" icon, you'll find only one: the army. "The U.S. Army wants you, and your friends, to accept the eCYBERMISSION challenge!" proclaims the site. But why? While the army disclaims any benefit except the opportunity to "give back to America, by helping youth learn more about the areas of science, math and technology," its personnel gain access to young-sters, serving "as CyberGuides, on-line experts who will communi-cate over the web, and Ambassadors who visit schools to promote the competition and award prizes to regional winners."

As the largest of the service branches, it's hardly shocking that the army tries to be all things to get all the people. Its lavish GoArmy.com Web site has a complete range of bells and whistles as well as links to plenty of other kid-friendly army Web pages, including those of the army Bull Riding & Rodeo team, the U.S. Army All-American Bowl (a nationally televised all-star high school football game), the Army Marksmanship Unit (which deploys to high schools, allowing kids to fire air rifles in exchange for personal information, like names and social security numbers), the army-sponsored Arena Football League, the Golden Knights parachute stunt team, the Takin' It to the Streets urban street team tour, army NASCAR auto racing, the army's National Hot Rod Association Top Fuel dragster and Pro Stock Bikes, and the army Auto Super Show.

The site, not surprisingly, is also loaded with ways for youths to interface with the army, including the ability, with a click of the mouse, to request an information packet; e-mail a question; enter a chatroom to talk with a recruiter; locate a flesh-and-blood recruiter or learn how to enlist. But perhaps the most powerful recruiting tool on the site is a page devoted to the America's Army's video game Web site.

Obviously, the video game and its official Web site are prime recruiting tools, but that doesn't stop the army from claiming oth-erwise. Its site explains it this way:

The game is designed to provide young adults and their influencers with virtual insights into entry level Soldier training, training in units and army operations so as to provide insights into what the army is like . . . Therefore, the game is designed to substitute virtual experiences for vicarious insights. It does this in an engaging format that takes advantage of young adults' broad use of the Internet for research and communication and their interest in games for entertainment and exploration.

Christopher Chambers, the deputy project director for America's Army, seems to have been equally disingenuous: "The army started this project as a means of 'connecting' with America. We look at the game as a first step in an information gathering process, not as a tool to bring in new recruits."

But if it isn't paying recruiting dividends, it would be hard to fathom why the army spent $19 million creating the game and continues to pour in $5 million per year to keep the game current. In fact, for a game specifically not designed for recruiting purposes, there's a whole lot of recruiting going on. For example, AmericasArmy.com offers links to the GoArmy.com recruiting Web site and advice on how to "initiate contact with an Army Recruiter." In the game, itself, there are, says the Web site,

> web links through which players can connect to the Army of One homepage. On GoArmy.com you can explore army career opportunities or contact a Recruiter . . . Players who request information AND reveal their nom-de-guerre to Recruiters may have their gaming records matched to their real-world identities for the purpose of facilitating career placement within the Army.

It's an approach that's paying off. While it's impossible to know how many people have enlisted as a direct result of the game, a 2004 study by the public relations research firm I-to-I, suggested America's Army was more effective than all other army branding efforts, combined, in exposing young people to the idea of a career in the military. According to Chambers, the America's Army "brand became so powerful that 29 percent of young people in the U.S. not only recognize the brand, but claim that the America's Army game is their primary source of information about the army." "What this tells us," he says, "is that we are having a very large impact on getting information about the army out to young people."

(MILITARY) CULTURE JAMRS

While the military's Web sites work their magic, BeNow, TRU, and other private contractors continue working with JAMRS—the DoD's "program for joint marketing communications and market research and studies"—to fill the ranks. JAMRS claims that it's only developing "public programs [to] help broaden people's understanding of Military Service as a career option." However, it also hires firms to engage in all sorts of not-for-public-consumption studies meant to "bolster the effectiveness of all the Services' recruiting and retention efforts." In other words, behind the scenes, the military is in a frantic search for weak points in the public's growing resistance to joining the armed services. Some of this research is impossible to learn about because access to the studies via the JAMRS Web portal is restricted. Should you attempt to examine the research, you are told that "access is currently limited to certain types of users"—none of which are you.

What we do know, however, is that JAMRS is focused on the following areas of interest in trying to bolster the all-voluntary military:

- **Hispanic Barriers to Enlistment.** A project to "identify the factors contributing to under-representation of Hispanic youth among military accessions" and "inform future strategies for increasing Hispanic representation among the branches of the Military."

- **College Drop Outs/Stop Outs Study.** A project aimed at gaining "a better understanding of what drives college students to . . . 'drop out' and determine how the Services can capitalize on this group of individuals (ages 18–24)."

- **Mothers' Attitude Study.** "This study gauges the target audience's (270 mothers of 10th- and 11th-graders) attitudes toward the Military and enlistment."

Eyebrows ought be raised over a Pentagon that is looking into ways to influence the mothers of teens to send their sons and daughters off to war—especially since JAMRS's survey found that "even those mothers who reported being in favor of military service were not overwhelmingly positive about their child joining the military." JAMRS also "identified messages that had a positive effect on different types of mothers" and found that if they could get "guidance counselors, teachers, and coaches" to lobby moms, they could make far greater recruitment inroads.

The military is no less eager to explore how the "elusive audience" of kids who drop out of school might be scooped up. The army's own internal manual for recruiters refers to the "college recruiting market" as "an excellent source of potential army enlistments due to the high percentage of students who drop out of college." It suggests to recruiters: "Focus on the freshman class because they will have the highest dropout rate. They will lack both the direction and funds to fully pursue their education."

Perhaps that's why, just before and after the end of the first semester of 2006, army recruiters and entertainment teams could be found trolling the campuses of Texas Christian University, the University of Louisville, Georgetown University, the University of South Dakota, the University of Illinois, Penn State University, Austin Peay State University, and Ohio State University (all of which received DoD dollars in 2006) as well as the Champs Sports Bowl Welcome Back to School Event and Capital 1 Bowl Welcome Back to School Event (both in Orlando, Florida) and the Chick-fil-A Bowl Welcome Back to School Event in Atlanta, Georgia.

Perhaps the most intriguing line of JAMRS's research, however, is the 2004 Moral Waiver Study, with the goal of better defining "relationships between pre-Service behaviors and subsequent Service success." What the JAMRS informational page doesn't make clear, but what might be better explained in the password-protected section of the site, is that a "moral character waiver" is the means by which potential recruits with criminal records are allowed to enlist in the military.

Interestingly, by February 2006, the *Baltimore Sun* observed that there was "a significant increase in the number of recruits with what the army terms 'serious criminal misconduct' in their background"—a category that included "aggravated assault, robbery, vehicular manslaughter, receiving stolen property and making terrorist threats." From 2004 to 2005, the number of those recruits had spiked by more than 54 percent, while alcohol and illegal drug waivers—reversing a four-year downward trend—had increased by over 13 percent. In June 2006, the *Chicago Sun-Times* reported that, under pressure to fill the ranks, the army had been enlisting increasing numbers of "recruits convicted of misdemeanor crimes, according to experts and military records." As the military's own data indicated, "The percentage of recruits entering the Army with waivers for misdemeanors and medical problems ha[d] more than doubled since 2001."

FUTURE SHOCK

JAMRS's partner Mullen Advertising works with it "on an array of marketing communications, planning, and strategic initiatives." One Mullen effort is the very unmilitary-sounding MyFuture.com, a slick Web site with information on such topics as living on your own, writing a cover letter, and finding a job. It even includes tips on dressing for success. ("Take extra time to look great.") MyFuture offers what seems like civilian career advice (albeit with some military images sprinkled throughout). You can, for instance, take its Work Interest Quiz in order to discover if you should "go to college or look for a job." However, this site, too, is naturally all about steering youngsters toward the armed forces. For example, when you take that quiz, you are prompted to ask your school guidance counselor "about taking the ASVAB Career Exploration Program if you'd like to know more about your aptitudes, values, and interests." Although it goes unmentioned, the acronym ASVAB actually stands for the Armed Services Vocational Aptitude Battery, a test developed during the Vietnam War, that has been called "the admissions and placement test for the US military" by the American Friends Service Committee. In fact, while the ASVAB Web site publicly claims that the test "assesses academic ability and predicts success in a wide variety of occupations," an army manual states that it is "specifically designed to provide recruiters with a source of pre-qualified leads."

When I took the quiz, I was told: "Based on your responses to the activities listed, here are the work styles that may be appropriate for you: Investigative [and] Artistic." Following up on my investigative aptitude, MyFuture.com offered eight civilian career suggestions, ranging from veterinarian to meteorologist. It also recommended eight military counterparts, including law enforcement and security specialist. For my artistic aptitude, MyFuture suggested that I "may like activities that: 'Allow [me] to be creative [and] Let [me] work according to [my] own rules.'" Apparently, there are eight military jobs that will allow me to stretch my imagination and do just what I want, artistically speaking. Who knew, for example, that the perfect move for an artistic free-thinker would be joining an organization based on authority and conformity—and then becoming a food service specialist?

TOMORROW'S MILITARY, TODAY?

Another Mullen Advertising–created site is aimed at a different population. TodaysMilitary.com announces that it "seeks to educate parents and other adults about the opportunities and benefits available to young people in the Military today." In JAMRS-speak, that means it's a "public site targeted at influencers."

TodaysMilitary.com is filled with information on financial incentives available to those who join the military and Web pages devoted to "what it's like" to be in the armed forces and how the military can "turn young diamonds in the rough into the finest force on the face of the earth." We learn that army basic training is "more than just pushups and mess halls." In fact, quite the opposite of a torture test, it's actually a "nine-week-long journey of self-discovery." Scanning through the pages, we even learn that in their off-time, military folk "go for walks . . . and they even shop for antiques" (which may account for some of the antiquities that seem to have gone missing from Iraq).

TodaysMilitary.com even takes the time to dispel "myths" like "People in the Military are not compensated as well as private sector workers." According to the site, they are—just don't tell it to the group of resentful marines who, in 2005, reportedly roughed up their highly paid mercenary counterparts in Iraq. "One Marine gets me on the ground and puts his knee in my back. Then I hear another Marine say, 'How does it feel to make that contractor money now?'" reported a former marine working as a "private security contractor" in the war zone. Armed private contractors in Iraq generally rake in $100,000 to $200,000 per year. That same year, under pressure from Congress, the Pentagon announced that it, too, would start paying out this type of cash. One caveat: You had to be dead.

FIGHTING SOLDIERS FROM THE SKY, FEARLESS MEN WHO JUMP AND [CENSORED] . . .

At such military Web sites, unpleasantries like death and danger typically go unmentioned. At TodaysMilitary.com, for instance, the only such allusion is on a Web page that coaches parents on ways to push their children to consider the military. Among the "tough issues" a child might raise is a simple fact, driven home nightly on the news: "It's dangerous." The site offers the following answer to parents: "There's no doubt that a military career isn't for everyone. But you and your young person may be surprised to learn that over 80 percent of military jobs are in non-combat operations . . . A military career is often what you make of it." Tell that to noncombat troops like the late corporal Holly Charette and her fellow sixteen casualties from a suicide car-bomb attack on a Marine Corps Civil Affairs team in Fallujah during the summer of 2005, or the hundreds upon hundreds of other troops in support roles who have found themselves directly in harm's way in Iraq. As a Voice of America article put it, "Increasingly, there is a fine line between combat and non-combat jobs, especially in a place like Iraq, where there is no front line, and any unit can find itself in a firefight at any moment."

ASSAULT AND (APTITUDE) BATTERY

In 2005, Major General Michael Rochelle, head of the Army Recruiting Command, stated: "Having access to 17- to 24-year-olds is very key to us. We would hope that every high school administrator would provide . . . lists [of student phone numbers and addresses] to us. They're terribly important for what we're trying to do." But as the *Nation* magazine editor Katrina vanden Heuvel wrote, some parents aren't so gung-ho: "The debacle in Iraq has made recruiting an impossibly difficult job and recruiters are sinking to new lows in the face of growing pressure to fulfill monthly quotas as well as fierce opposition from parents who don't support the President's botched Iraq war mission."

One of the military's new *lows* brings us back to the subject of the ASVAB and the methods of the Vietnam era. Faced then with the need for expendable troops, Secretary of Defense Robert McNamara instituted an unholy coupling of the War on Poverty and the War in Vietnam: Project 100,000. It called for the military, each year, to admit 100,000 men who had failed its qualifying exam. Those who failed to meet minimal mental standards, men McNamara called the "subterranean poor," would be offered an education and training that would be useful upon their return to civilian life. Instead of acquiring skills useful for the civilian job market, however, "McNamara's moron corps" (as they came to be known within the military) were trained for combat at markedly elevated levels, were disproportionately sent to Vietnam, and had double the death rate of American forces as a whole. Today, the

Pentagon is increasingly turning to recruits scoring in the lowest level on the Armed Services Vocational Aptitude Battery, accepting enlistees who would have been rejected a few years earlier.

STOP AND SELL WITH FLOWERS

Other lows are unabashedly showcased in the U.S. Army Recruiting Command's September 2004 *School Recruiting Program Handbook*. In the guide, designed to provide recruiters with "a single-source document" for making inroads with the teen set, the military lays out its plan of attack for schools. According to the handbook, recruiters "must convince school officials that they actually have their students' best interests in mind." To do this, the manual suggests engaging in old-fashion brownnosing. Tips include:

> Never forget to ask school officials if there is anything you can do for them.

> Be so helpful and so much a part of the school scene that you are in constant demand.

> Cultivate coaches, librarians, administrative staff, and teachers . . . By directing your efforts toward other faculty members you may be able to obtain information necessary to effectively communicate with students.

The guide also recommends that recruiters become vital assets to a high school's sports programs by joining the coaching corps. Other methods include acts of bribery—using flowers and doughnuts to sweeten the deal.

> Your goal is to develop as many COIs [Centers of Influence] as possible. Don't forget the administrative staff . . . [e]stablish and maintain rapport . . . [and] have something to give them (pen, calendar, cup, donuts, etc.,) and always remember secretary's week with a card or flowers.

Deliver donuts and coffee for the faculty once a month. This will
help in scheduling class-room presentations.

The army even recommends *enlisting* kids in the recruitment
effort. While the handbook concedes that "some influential students like the school president or captain of the football team may
not enlist . . . they can and will provide you with referrals on who
will enlist." To this end, it exhorts recruiters to identify "student
influencers," lobby them, and even hand out promotional items to
win them over.

Obviously the Pentagon thinks that America's youth (and its
teachers and parents) couldn't really pass a basic intelligence test
or can be bought off with cheap trinkets. Its Web sites downplay
danger and offer bloodless scenarios filled with adventure and
heroism that don't square with news coming home from Iraq. Its
methods are just one more example of how the Complex works to
sustain itself by whatever means necessary.

15

THE MILITARIZATION OF MYSPACE

Those young years can be hard ones. The acne, the awkwardness, the angst. That may be one reason why, if you're between thirteen and thirty, you're probably already making "friends" in the cozy cyberconfines of MySpace.com, the social-networking Web site which bills itself as "an online community that lets you meet your friends' friends." At MySpace, each user can create a customized Web page or "profile," upload photos (only from your best angle and Photoshopped to the hilt), blog around the clock, and—most important of all—court those "friends."

In an eerie reflection of the very world that many MySpace scenesters undoubtedly plunge into cyberspace to avoid, *the* measure of success at the site is how much you can increase your page's popularity. You do this by posting attention-grabbing content, breathlessly soliciting other users, putting up provocative pictures to attract attention, sending out "bulletins" to your existing "friends" and asking them to "whore" you out to their list of friends. With its multimillions of "friends" to garner, the site is wildly popular—and not just for insecure teens either.

MySpace has become a magnet for those who want, for one reason or another, to draw young eyeballs (and often pocketbooks). Colleges, corporate products like Toyota's Yaris and the

Honda Element, even fictional characters like Ricky Bobby from the movie *Talladega Nights* or fast-food outlet Wendy's minimalist cartoon pitchman Smart have already gotten into the MySpace act.

Early in August 2006, the site hit a major milestone: 100 million profiles. Even including those corporate-sponsored sites and fictional pages, that's still a whole lot of would-be *friends*. That same month, *Fortune* magazine reported that MySpace, bought up by FOX News mogul Rupert Murdoch in 2005 as part of a $580-million deal, "passed Google in terms of traffic" and now ranks second only to Yahoo in page views with one billion daily. Already "home to 2.2 million bands, 8,000 comedians, thousands of filmmakers, and millions of striving, attention-starved wannabes," the magazine reported that, on a "typical day," it signs up 230,000 new users. In fact, as of April 2007, according to Internet ratings agency Hitwise, MySpace received 79.7 percent of all visits to social-networking Web sites.

Obviously, the site has shown special skill in *recruiting* people since its launch in 2003. In the same years that MySpace has become an Internet superpower, the U.S. armed forces have sustained substantial losses. Little wonder then, with 80 percent of MySpace users reporting they're over eighteen years old, that the military set its sights on occupying some virginal virtual territory in search of fresh-faced recruits who might be thrown into the Afghan and Iraqi breaches.

In February 2006, the Marine Corps launched its MySpace profile. A thoroughly predictable page, it boasted a streaming video that might best be termed boot-camp-on-speed and the usual downloadable desktop wallpapers, mainly Marine Corps "anchor and globe" emblems or photos of World War II vintage marines. By July 2006, according to an Associated Press report, a modest "430 people ha[d] asked to contact a Marine recruiter through the site . . . including some 170 who [we]re considered 'leads' or prospective Marine recruits." By September of the next year, the marines had nearly 56,200 MySpace "friends" endorsing their page, less than half the "friends" of Senator Hillary Clinton and over 6,000 fewer than the Yaris, but a respectable number nonetheless.

Not to be left out, in August 2006, the air force launched its own page. Colonel Brian Madtes, the Air Force Recruiting Service's strategic communications director, was blunt about the reasons in an interview with the air force's own news agency: "In order to reach young men and women today, we need to be in tune and engaged in their circles."

One-upping the marines, the air force also launched a cross-promotional effort with the FOX network television show *Prison Break*. Visitors to its MySpace profile page were offered five slick "rough cuts" of air force commercials on which to vote their preferences. The winning ad ran during the September 18, 2006, episode of the prison-escape drama. But the next day, in an abrupt about-face, the air force shut down its MySpace page over "concerns that association with inappropriate content might damage the service's reputation." As Madtes told the *Air Force Times,* "The danger with MySpace is we got to the point where we weren't real comfortable with the potential for inappropriate content to be posted [on the page of] a friend of a friend. We didn't want to be associated with that and tarnish our reputation."

In February 2006, the army also expressed reservations over MySpace and canceled an advertising contract with the site after just one month, due to reports of "child predators approaching youths via the site." Then attorney general Alberto R. Gonzales had only recently called attention to an incident in which a man "used 'MySpace.com' to lure an 11-year-old girl into having illicit sexual relations," and the House of Representatives passed a measure to ban MySpace.com and other social-networking sites from schools and libraries by a lopsided 410-to-15 vote. Then, during the summer of 2006, an army sergeant, based in Fort Drum, New York, was caught in a sting operation soliciting a sheriff's detective, posing as a fifteen-year-old girl on MySpace, for sex. He pleaded guilty to "criminal solicitation and attempted rape in the third degree."

In the end, despite these developments, the army decided to embrace MySpace in a bigger way. According to Louise W. Eaton, the service's advertising media and Web chief, MySpace production teams worked with army Web designers and a team from McCann Erickson, the army's ad agency, to create an interactive site complete with all the necessary bells and whistles along with "several ways to contact a recruiter." While the army's designers are primarily after the eyeballs of "enlistment prospects" between the ages of seventeen and twenty-four, she recognized that a younger set might also be taking a look. "It's alright for younger people to see it, it's not propaganda," she commented.

According to Eaton, the army's MySpace.com profile page is entirely devoted to shuttling people to the official GoArmy Web site. Taking a page from then secretary of defense Donald Rumsfeld's book, she defined success—in this case online, not in Iraq—in terms of "metrics." For her, three were key: page views, people who contact the army for "more information," and traffic to GoArmy.com. "We'll be very interested to see how many people register as our friend," she confessed, suggesting that she expected them to be "very, very, very many in number." In fact, by September 2007, the army had nearly 26,000 friends—far fewer than the unofficial Noam Chomsky page's over 51,000, but still a substantial figure. The army's MySpace effort had lured them in with a page rich with features: a discussion board, an interactive virtual army "guide," wallpapers, photos, video clips, links to the America's Army videogame and GoArmy.com as well as a "Contact the Army" link.

The army's eyes were also on "the blogosphere." Eaton noted that "many, many military people unofficially participate and we're studying that and trying to figure out where to go with that." By November 2007, the army decided that one way to go was the "G.I. Jill" blog—an official army site devoted to Sgt. Jill Stevens, a member of the Utah, National Guard's 211th Aviation Regiment, the reigning Miss Utah, and a contestant in the 2008 Miss America pageant. In addition to regular postings written by Stevens, and links to the army's own Web site, the blog even offered a downloadable pin-up of the sergeant, all made-up, seated on the hood of a military vehicle, cradling her automatic rifle with her crown sitting beside her.

The Pentagon's new focus on finding "friends" on social-networking sites is a symptom of how hard-pressed its officials really are—as the military continues to invade new media territory, from text messaging to Pentagon podcasting. What all this means is that sexual predators aren't the only ones trolling the Internet for young bodies. MySpace claims to be taking steps to safeguard children from a certain type of cyberstalker. But what kind of "friend" looks to enlist you in a potentially life-threatening enterprise, like the war in Iraq, already considered a catastrophe by most Americans?

The role of "friendly" MySpace.com, taking a desperate military's money to target hordes of young friends searching for popularity online, is troubling. It's also typical of the business side of the Complex, because it's the civilian firms that allow the military to function as it does. In the case of MySpace, the company is thoroughly involved in producing the army's page and will, says Eaton, be "doing the daily maintenance" on it.

If bios at the site are to be believed, there are young Iraqis on MySpace. What if an American kid with an Iraqi MySpace "friend" checks in with that friendly Marine Corps recruiter, enlists, and is sent to Iraq, where his/her MySpace military "friend" orders him/her to kill the former? What then?

16

THE DIRTY DOZEN

Even in the midst of the disastrous war in Iraq, military recruiting has been marked by upbeat pronouncements from the Department of Defense, claims of success by the White House, and periodic press reports touting the military's ability to meet its wo/manpower goals. But the armed forces only found success through a fundamental "transformation," and not the one former defense secretary Donald Rumsfeld had in mind before the invasion of Iraq. While his long-standing goal of transforming the planet's most powerful military into its highest-tech, most agile, most futuristic fighting force "melted away," according to the *Washington Post's* David von Drehle, the very makeup of the armed forces has, indeed, been mutating, even if in an unexpected fashion.

In 2005, despite spending at least $16,000 in promotional costs for each soldier it managed to sign up, the Pentagon failed to meet its recruiting goals. After that, recruitment methods were ramped up in twelve critical areas, some of which turned the old army adline, "Be All That You Can Be," into material for late-night talk show punch lines.

1. HARD SELL

When not trolling for potential soldiers via video games, Web sites, MySpace.com, and text messaging, the armed forces began mobilizing recruiters who used old-fashioned hard-sell tactics to cajole impressionable teens into enlisting. In 2006, one New Jersey mother told her local newspaper about the army's persistence in targeting her seventeen-year-old daughter. When the mother finally asked the army to stop calling, the recruiter argued vigorously against it. The mother, who otherwise praised the military, was aghast at the recruiter's tactics. "That's what frightened and enraged me. This military person telling me that I have no rights over my child."

Teens have also been subject to military advertising and high-pressure tactics at school. In 2006, the *Boston Globe* wrote that recruiters were now setting up booths in "cafeterias in high schools across the nation." That same year, the *State Journal-Register* of Springfield, Illinois, reported that local recruiters were "visiting each school about every three to four weeks." At one school, administrators were forced to "clam[p] down on aggressive recruiters"; at least one was barred from ever returning to campus.

2. GREEN TO GRAY

The military has always filled its rolls primarily by targeting the young, but under the pressure of Iraq the "old" made it into its sights as well. In 2005, the Army Reserve increased its maximum enlistment age from thirty-five to forty; then, later that year, to forty-two. The next year, regular army green went grayer with a similar two-step increase that boosted active-duty enlistment eligibility to forty-two years.

3. BACK-DOOR DRAFT

Another group of old-timers was also targeted by the military: the Marine Corps Individual Ready Reserve (IRR)—troops who had left active-duty status and transitioned back into civilian life. In August 2006, the marines announced that they would begin making up for a shortage of volunteers by "dipping into [this] rarely used pool of troops to fill growing personnel gaps in units scheduled to deploy in coming months." As the *Boston Globe* noted, it was the first time since the invasion of Iraq three years before that marine commanders had "taken the extraordinary step of drafting back into uniform those who have left the ranks." For its part, according to a 2006 CBS News report, the army had "called to active duty" approximately fourteen thousand soldiers on IRR status from the Army Reserve since March 2003, and over half were deployed to Iraq.

4. RUBBER-STAMP PROMOTIONS

In 2006, the army admitted that, to maintain desperately needed numbers, it was sacrificing almost any measure of quality when it came to its officer corps. According to 2005 Pentagon figures, 97 percent of all eligible captains were being promoted to major—a significant jump from the already historically high average of 70 to 80 percent. "The problem here is that you're not knocking off the bottom 20 percent," one high-ranking army officer at the Pentagon told the *Los Angeles Times*. "Basically, if you haven't been court-martialed, you're going to be promoted to major." Despite near-guaranteed promotions, *USA Today* reported, in 2007, that the army faced not only "an impending shortage of entry-level officers—lieutenants," but also their superiors. "We are short about 3,000 midgrade officers, particularly majors, and we will be for the next several years," acknowledged army spokesman Lieutenant Colonel Bryan Hilferty. In fact, an annual shortage of three thousand midlevel officers is expected through 2013.

5. FOREIGN LEGION

Testifying before the Senate Armed Services Committee, in July 2006, Undersecretary of Defense for Personnel and Readiness

David S. C. Chu listed a series of inducements then being offered to get foreigners to risk life and limb for Uncle Sam. These included: "President Bush's executive order allowing non-citizens to apply for citizenship after only one day of active-duty military service"; a streamlined application process for service members; and the elimination of "all application fees for non-citizens in the military."

While noting that approximately forty thousand noncitizens were already serving in the armed forces, Chu offered his own solution to the immigration crisis in the United States. With the services denied the possibility of a draft, he made a pitch for creating a true foreign legion from the "estimated 50,000 to 65,000 undocumented alien young adults who entered the U.S. at an early age." Chu then talked up legislation like the DREAM Act, which would give illegal aliens the opportunity to, among other options, join the military as a vehicle to conditional permanent resident status.

In addition to proposing a possible source of undocumented cannon fodder that might prove less disturbing to Americans than their own sons and daughters, Chu noted that the "military also has initiated several new programs, including opportunities for those with language skills, which may hold particular appeal for noncitizens." Just in case noncitizens aren't thrilled by the chance to serve, the army promises expedited citizenship, quick advancement, and a host of other perks—including a small boatload of cash. In addition to "foreign language proficiency pay while on active duty," those willing to use their "Middle-Eastern language skills and join the U.S. Army as a Translator Aide . . . in Iraq and Afghanistan" were to receive enlistment bonuses of $10,000—a sizable sum given yearly per capita incomes in the countries being targeted ($800–$2,000 a year).

6. MERCENARY MILITARY

To staunch its recruitment wounds, the military has also enhanced its offers—in the form of "more financial incentives." In some cases, this can mean enlistment bonuses as high as $40,000 for documented but poor Americans looking to put themselves in harm's way for three years as army infantrymen or explosive ordnance disposal specialists—markedly more than per capita yearly income in 2005 for African Americans ($16,874), Hispanics ($14,483), and non-Hispanic whites ($28,946). Even those not interested in volunteering for the most dangerous jobs can reap big bonuses. In the summer of 2007, in a bid to rapidly shove drastically needed recruits into the breach, the army increased its enlistment bonus to $20,000 for those willing to sign on for a "Quick Ship" provision that slashes the usual forty-day delay between sign-up and basic training. This was followed, in November, by a pilot program offering recruits $45,000 toward buying a house or starting a business upon completion of their military service.

According to the Associated Press, the army was also doling out fistfuls of dollars for reenlistees—in 2006, "an average bonus of $14,000, to eligible soldiers, for a total of $610 million in extra payments." By the next year, the army had raised the figure to $33,500 for certain critical specialities, with some soldiers eligible for lump-sum payments of up to $25,000. In July 2006, Major Jerry Morgan, who runs the Marines Selective Reenlistment Bonus Program, told *Stars and Stripes* that "the maximum bonus" had been raised to $60,000 for those serving in five critical military occupational specialties. Less than a year later, some Marine Corps bonuses reached the $80,000 level. Add to these sums promised benefits of up to $71,424 and $23,292 for active duty and reserve personnel to "help pay for college," and you've got a potentially life-changing bribe, provided you still have a life when that college acceptance finally comes through.

7. ABUSE OF POWER

More recruiters waving more money has its pitfalls, though. In 2005, amid a swirl of complaints as recruiters struggled to meet monthly goals (on the part of some by offering tips to potential enlistees on how to pass drug tests), the army suspended all recruiting activities for a one-day nationwide "stand down" to reexamine its methods and retrain its personnel. No wonder. A Government Accountability Office report later indicated that "between fiscal years 2004 and 2005, allegations and service-identified incidents of recruiter wrongdoing increased, collectively, from 4,400 cases to 6,500 cases; substantiated cases increased from just over 400 to almost 630 cases; and criminal violations more than doubled from just over 30 to almost 70 cases."

In August 2006, it was also revealed that "more than 100 young women who expressed an interest in joining the military in the past year were preyed upon sexually by their recruiters." According to one of the victim's lawyers, a recruiter "said to her, outright, if you want to join the marines, you have to have sex with me. She was a virgin. She was 17 years old." Another teenage victim spelled out the situation quite clearly, "The recruiter had all the power. He had the uniform. He had my future. I trusted him."

8. CIVILIAN HEADHUNTERS

Not surprisingly, given an administration that never saw anything it couldn't imagine privatizing, the private headhunter also joined the military recruitment landscape. According to Renae Merle of the *Washington Post,* as part of a pilot program that began in 2002, two Virginia-based companies, Serco (which received $21.8 million from the DoD in 2006) and MPRI (whose parent company, L-3 Communications, took home $5.2 billion from the Pentagon in 2006), had "more than 400 recruiters assigned across the country, and . . . signed up more than 15,000 soldiers." They were, she reported, "paid about $5,700 per recruit."

While these companies took in Pentagon dollars, the private recruiters themselves received cash bonuses, free gas cards, and suede jackets—and could augment their base salary by about $30,000 a year if they successfully shuttled large numbers of potential recruits into the armed forces. As has been true with the DoD's use of private contractors in all sorts of roles in recent years, this step drew some congressional ire. According to Representative

Jan Schakowsky (D-Illinois), "The use of contractors for this sensitive purpose, dealing with the lives of young people, is trouble-
some." She was particularly worried by the lack of oversight, while
an army report that recommended continuing the $170-million
program also noted that the civilian headhunters "enlisted a lower
quality of recruit."

Despite these quality-control issues, in August 2006, army officials announced that they had awarded MPRI a base contract of
$11,196,996, potentially worth over $34 million in total, for "recruiting services to . . . be performed at any of the Army's 1,700 recruiting
stations nationwide." And then, in February 2007, they awarded
MPRI yet another multimillion-dollar recruiting-services contract.

9. HOW LOW CAN YOU GO?

In this period, lowered standards have been typical of more than
the privateers. Even Undersecretary of Defense Chu admitted that
nearly 40 percent of all military recruits in 2006 scored in the
bottom half of the armed forces' own aptitude test. Other how-
low-can-you-go indicators of the military desperation regularly
surfaced in news reports. For example, in 2005, the New York Times
reported that two Ohio recruiters were quick to sign up a recruit
"fresh from a three-week commitment in a psychiatric ward . . .
even after the man's parents told them he had bipolar disorder—a
diagnosis that would disqualify him." After senior officers found
out, the mentally ill man's enlistment was canceled, but in "interviews with more than two dozen recruiters in 10 states," the Times
heard others talk of "concealing mental-health histories and
police records," among other illicit practices.

In May 2006, the Oregonian reported that army recruiters, using
hard-sell tactics and offering thousands of dollars in enlistment
bonus money, signed up an autistic teenager "for the army's most
dangerous job: cavalry scout." The boy, who had been enrolled in
"special education classes since preschool" and through "a special
program for disabled workers . . . ha[d] a part-time job scrubbing
toilets and dumping trash," didn't even know the United States

was at war in Iraq until his parents explained it to him after he was approached by a recruiter. Only following a flurry of negative publicity did the army announce that it would release the autistic teen from his enlistment obligation.

10. ARMED AND CONSIDERED DANGEROUS

The Pentagon's 2004 "Moral Waiver Study" led the way to opening recruitment doors to potential enlistees with criminal records, and one beneficiary of the army's moral-waiver policies did gain a certain notoriety. After Steven D. Green, who served in the army's 101st Airborne Division, was charged in a rape and quadruple murder in Mahmudiyah, Iraq, it was disclosed that he had been "a high-school dropout from a broken home who enlisted to get some direction in his life, yet was sent home early because of an 'anti-social personality disorder.'" Eli Flyer, a former Pentagon senior military analyst and specialist on "the relationship between military recruiting and military misconduct," told *Harper's* magazine that Green had actually "enlisted with a moral waiver for at least two drug- or alcohol-related offenses. He committed a third alcohol-related offense just before enlistment, which led to jail time, though this offense may not have been known to the Army when he enlisted."

With Green in jail awaiting trial, the *Houston Chronicle* reported, in August 2006, that army recruiters were trolling the outskirts of a Dallas-area job fair for ex-convicts. "We're looking for high school graduates with no more than one felony on their record," one recruiter said. The army even considered the incarcerated fit for military service—in one case recruiting in a "youth prison" in Ogden, Utah. Although Steven Price asked to see a recruiter while still inside and was "barely 17 when he enlisted," in January 2005, his divorced parents claimed to a Utah TV reporter that "recruiters used false promises and forged documents to enlist him." While confusion existed about whether the boy's mother actually signed a parental consent form allowing her son to enlist, his father, according to the reporter, "apparently wasn't even at the signing, but his name [wa]s on the form too."

11. GANG WARFARE

According to the *Chicago Sun-Times*, law enforcement officials have reported that the military is now "allowing more applicants with gang tattoos because they are under the gun to keep enlistment up." They also note that "gang activity may be rising among soldiers." The paper was provided with "photos of military buildings and equipment in Iraq that were vandalized with graffiti of gangs based in Chicago, Los Angeles and other cities."

The *Sun-Times* also reported that a gang member facing federal charges of murder and robbery enlisted in the Marine Corps in 2006, while "free on bond—and was preparing to ship out to boot camp when Marine officials . . . discovered he was under indictment." While this particular recruit was eventually booted from the corps, a Milwaukee police detective and army veteran, who served on the federal drug and gang task force that arrested the would-be marine, noted that other "gang-bangers are going over to Iraq and sending weapons back . . . gang members are getting access to military training and weapons."

A report by the army's Criminal Investigation Command noted, wrote *Stars and Stripes*, that "the number of gang-related crimes involving soldiers and their families nearly tripled from fiscal 2005 to fiscal 2006," while suspected gang activities or crimes were reported at eighteen separate bases. Other federal agencies have also taken notice. In 2007, a Federal Bureau of Investigation (FBI) National Gang Intelligence Center report noted that "gang-related activity in the military is increasing and poses a threat to law enforcement officials and national security."

12. TRADING DESERT CAMO FOR WHITE SHEETS

Another type of "gang" member has also begun to multiply within the military, evidently thanks to lowered recruitment standards and an increasing urge by recruiters to look the other way. In July 2006, a study by the Southern Poverty Law Center, which tracks racist and right-wing militia groups, found that—due to pressing manpower concerns—"large numbers of neo-Nazis and skinhead extremists" are now serving in the military. "Recruiters are knowingly allowing neo-Nazis and white supremacists to join the armed forces and commanders don't remove them from the military even after we positively identify them as extremists or gang members," said Scott Barfield, a Defense Department investigator quoted in the report.

The *New York Times* noted that the neo-Nazi magazine *Resistance* was actually recruiting for the U.S. military, "urg[ing] skinheads to join the Army and insist on being assigned to light infantry units." As the magazine explained, "The coming race war and the ethnic cleansing to follow will be very much an infantryman's war . . . It will be house-to-house . . . until your town or city is cleared and the alien races are driven into the countryside where they can be hunted down and 'cleansed.'"

Apparently, the recruiting push has worked. Barfield reported that he and other investigators identified a network of neo-Nazi active-duty army and marine personnel spread across five military installations in five states. "They're communicating with each other about weapons, about recruiting, about keeping their identities secret, about organizing within the military." Little wonder that "Aryan Nations graffiti" is now apparently competing for space among American inner-city gang graffiti in Iraq.

FORCE TRANSFORMATION

When the American war in Vietnam finally ground to a halt, the U.S. military was in a state of near disintegration. Uniformed leaders vowed never again to allow it to be degraded to such a point. A generation later, as the ever-worsening wars in Iraq and Afghanistan grind on, an overstretched army and Marine Corps, at a remarkable cost in dollars, effort, and lowered standards, are fighting to maintain their numbers. As a result, U.S. ground forces are increasingly made up of a motley mix of underage teens, old-timers, foreign fighters, gangbangers, neo-Nazis, ex-cons, inferior

officers and near-mercenary troops, lured in or kept in uniform
through big payouts and promises.

   In the years when the breakdown was occurring during the
Vietnam War, American troops began to scrawl *UUUU* on their helmet liners—an abbreviation that stood for "the unwilling, led by
the unqualified, doing the unnecessary for the ungrateful." The
U.S. ground forces of this era may come to increasingly resemble
the collapsing military of the Vietnam War, the band of fictional
criminal misfits sent behind enemy lines in the classic Vietnam-
era film (though it was set in World War II) *The Dirty Dozen*,
or even the janissaries of the old Ottoman Empire. In fact, a
new all-volunteer generation of UUUUs may be emerging: the
underachieving, unintelligent, unsound, unhinged, unacceptable,
unhealthy, undesirable, uncivil, and even un-American, all led by
the unqualified, doing the unnecessary for the ungrateful. This
wasn't the new military Donald Rumsfeld was promising all those
years, but there's no denying the depth of the transformation.

PART VI

THE MAD, MAD WORLD OF THE MILITARY

By now the military-telecom complex is old hat to you—but did you know that you could tie it to the military-golf complex? (Yes, Virginia, there is a military-golf complex.) You could also tie it to the military's media empire and to the highest-paid athlete in the world. On July 4, 2007, Randall Stephenson, the chairman of AT&T (which received more than $230 million from the DoD in 2006) took to the links for a pretournament practice round with an army sergeant, an air force sergeant, and sports legend and General Motors (over $87 million from the Pentagon in 2006) pitchman Tiger Woods. Former president George H. W. Bush and Barbara Bush even joined the foursome for a few holes. This was all dutifully reported by the American Forces Press Service—the Pentagon's very own propaganda arm—because Woods donated thirty thousand tournament tickets to military personnel to attend the Earl Woods Memorial Pro-Am, a tournament named after Woods's late Green Beret father.

While it didn't take place on one of the scores of military golf courses located worldwide today, Woods did mention the existence of this vast complex of links, explaining that he had "played a lot of military facilities around the country." He also claimed that if he hadn't been making $100 million per year, he'd "probably end up going into the military." " I don't know what branch," said Woods, "but I'd certainly want to be in the special operations community."

Instead of black ops, however, Woods took part in photo ops that found their way onto numerous military Web sites from Americasupportsyou.com to Defenselink.mil. These "news" outlets and the military's golf courses are just part of a mad, mad world the military has made—a world of tax-subsidized golf courses, propaganda posing as news, and press-the-flesh conferences brimming with generals as well as plenty of barbequed pork.

You couldn't make this stuff up if you tried.

17

THE PENTAGON
GOES GOLFING

Back in 1975, Senator William Proxmire (D-Wisconsin) decried the fact that the Department of Defense spent nearly $14 million each year to maintain and operate 300 military-run golf courses scattered across the globe. In 1996, the weekly television series *America's Defense Monitor* noted that "Pentagon elites and high government officials [were still] tee-ing off at taxpayer expense" at some "234 golf courses maintained by the U.S. armed forces worldwide." In the intervening twenty-one years, despite a modest decrease in the number of military golf courses, not much had changed. The military was still out on the links.

Today, the military claims to operate a mere 172 golf courses worldwide, suggesting that over thirty years after Proxmire's criticisms, a modicum of reform has taken place. Don't believe it. In actuality, the military has cooked the books. For example, the Department of Defense reported that the U.S. Air Force operates 68 courses. A closer examination indicates that the DoD counts the 3 separate golf courses, a total of fifty-four holes, at Andrews Air Force Base in Washington, D.C., as 1 course. The same is true for the navy, which claims 37 courses (including facilities in Guam, Italy, and Spain) but counts, for example, its Admiral Baker Golf Course in San Diego, which boasts 2 eighteen-hole courses, as a single unit. Similarly, while the DoD claims that the army operates

56 golf facilities, it appears that this translates into no fewer than 68 actual courses, stretching from the U.S. to Germany, Japan, and South Korea.

Moreover, some military golf facilities are mysteriously missing from all lists. In 2005, according to the Pentagon, the U.S. military operated courses on twenty-five bases overseas. (See accompanying chart.)

Air Force	Marine Corps	Army	Navy
Ramstein, Germany	Butler, Okinawa	Baumholder, Germany	Guam
Aviano, Italy		Edelweiss, Germany	Naples, Italy
Misawa, Japan		Heidelberg, Germany	Atsugi, Japan
Yokota, Japan		Kitzingen, Germany	Rota, Spain
Kunsan, Republic of Korea		Stuttgart, Germany	
Osan, Republic of Korea		Wiesbaden, Germany	
Kadena, Okinawa		Camp Zama, Japan	
Incirlik, Turkey		Camp Casey, Republic of Korea	
Lakenheath, United Kingdom		Camp Red Cloud, Republic of Korea	
		Camp Walker, Republic of Korea	
		Sungnam, Republic of Korea	

A closer look, however, indicates that the military apparently forgot about some of its golf courses—especially those in unsavory or unmentionable locales. Take the unlisted eighteen-hole golf course—where hot-pink balls are used so as not to lose them in the barren terrain—at the U.S. naval base at Guantánamo Bay, Cuba. Also absent is the army's Tournament Players Club, a golf course built, in 2003, by army personnel in Mosul, Iraq. Another forgotten course can be found in the Republic of the Marshall Islands, at Kwajalein, a little-discussed island filled with missile and rocket launchers and radar equipment that serves as the home of the U.S. Army's Ronald Reagan Ballistic Missile Defense Test Site. Similarly unlisted is a nine-hole golf course located on the shadowy island of Diego Garcia, a British Indian Ocean Territory occupied by the U.S. military and long suspected as the site of one of the CIA's post-9/11 secret "ghost" prisons. But even courses not operating on secret sites, in war zones, or near prisons and possible torture centers have been conveniently lost. For example, while the Pentagon lists the navy's Admiral Nimitz Golf Course in Barrigada, Guam, in its inventory of overseas courses, it seems to have skipped Andersen Air Force Base's eighteen-hole Palm Tree Golf Course, also on the island. And you'd think the Pentagon would be proud of the USAF's island links; after all, it was the runner-up, in 2002, for the title of "Guam's Most Beautiful Golf Course."

Whatever the true number of the military's courses, at least some of them are distinctly sprucing up their grounds. Take the Eaglewood Golf Courses at Langley Air Force Base in Virginia. In 2004, the Pentagon paid out more than $352,000 to George Golf Design to refurbish its two courses (known as "the Raptor" and "the Eagle"). George Golf Design considerately worked on the courses one at a time, so that local duffers would not be left linkless. This was of critical importance since if both courses were out of commission, Virginia would have been left with only nine military golf facilities (navy, five; army, three; Marine Corps, one) with a total of fourteen courses.

Even though the military operates so many courses, apparently these still aren't enough to satisfy the insatiable golfing appetites of the armed forces—at least judging by the number of golf resorts to which the Pentagon paid out American tax dollars in 2004. For instance, the Del Lago Golf Resort and Conference Center, in San Antonio, Texas, which offers an "18-hole championship golf course home to some of the region's most challenging and beautiful holes," received over $19,000, and the Lakeview Golf Resort and Spa in Morgantown, West Virginia, which boasts "two championship golf courses," received $16,416 from the army in 2004. When asked what exactly the army was up to at Lakeview, a resort spokesperson declined to "disclose any information" and stated that she was "unable to confirm activities" of the military at the resort if, in fact, they occurred at all.

At the Arizona Golf Resort and Conference Center in Mesa, Arizona, which boasts "fine accommodations, great dining and a host of amenities, including a championship golf course, surrounded by beautifully maintained grounds," the army dropped a cool $48,620 in 2004. That resort wasn't, however, the top recipient of military funds among *Arizona* golf resorts. That year, according to DoD documents, the U.S. Army paid $71,614 to the Arizona Golf Resort—located in sunny Riyadh, Saudi Arabia.

A Saudi homage to the American Southwest that claims to offer the "only residential western expatriate golf resort in Riyadh with activities for all ages," the resort actually boasts an entire entertainment complex, complete with a water-slide-enhanced megapool, gym, bowling alley, horse stables, roller hockey rink, arcade, amphitheater, restaurant, and even a cappuccino bar—not to mention the golf course and a driving range. It's the perfect spot, in the so-called arc of instability, for military folks to play a few rounds with other Westerners. For those in the Persian Gulf who prefer their links on a smaller scale, there are also miniature golf courses at such military bases as Ahmed Al Jaber Air Base and at Camp Doha, both located in Kuwait, Balad Air Base in Iraq, and the air force's base at Eskan Village, near Riyadh Air Base, in Saudi Arabia.

But minigolf isn't the only activity for duffers stationed at Eskan. In 2002, the U.S. General Accounting Office investigated "seemingly unneeded expenditures" by the military and found that $5,333 had been spent on "golf passes" for folks from Eskan Village. In fact, the GAO reported: "Air Force units purchased several golf items during their deployments to Southwest Asia that included a golf cart for $35,000, a corporate golf membership at $16,000 . . . and a golf club/bag set costing nearly $1,500." The military's ardent love affair with golf carts hardly ended with that $35,000 model. In 2004 alone, according to the Pentagon's own documents, the DoD paid $6,860 to Golf Car Company, $6,900 to Golf Cars of Riverside, $9,322 to Golf Cars of Louisiana, $16,741 to Southern Golf Car, and a whopping $37,964 to Golf Car Specialties. Similarly, in 2006, two golf cart concerns were paid a combined $58,644 by the DoD, while a German golf-equipment supplier, Continental Golf Associates, received more than $88,000 from the Pentagon.

Despite base closures and the work of committed environmental and community groups, which have thinned out some of the military's links, the Department of Defense continues to exhibit an obsession with golf, golf carts, and, above all, golf courses. Apologists, both within and outside of the military, often counter criticisms of DoD golf expenditures by claiming that military golf courses are not simply a drain on taxpayer money but revenue earners, through greens fees. They, however, never make mention of the fact that these facilities are located on public land and pay no taxes; that they require funds for security; and that in all likelihood the public pays for the roads, water, and electric lines that service the courses—sore points raised by former Arizona senator Dennis DeConcini in the mid-1990s when Andrews Air Force Base was sinking $5.1 million into its third course. (If the DoD really wanted to raise revenues, it would sell its courses. For example, the army's Garmisch, Kornwestheim, and Heidelberg golf courses in Germany are worth, says the DoD, $6.6 million, $13.3 million, and $16.5 million, respectively, while the DoD's Sungnam golf course in the Republic of Korea is reportedly valued at $26 million.)

Such a defense also fails to address why the Pentagon is in the golf course business in the first place. According to its officially stated mission, the DoD engages in war-fighting, humanitarian, peacekeeping, evacuation, and homeland-security missions and, says the Pentagon, provides "the military forces needed to deter war and to protect the security of the United States. Everything we do supports that primary mission." How, exactly, golf courses ensure that primary mission is a little murky, especially since the United States has more than 8,100 public courses and over 3,500 semiprivate courses (that allow some access to nonmembers). A more apt explanation is the fact that when it comes to golf, like much else, the Pentagon does what it wants, no matter who gets tee'd off.

18

PATRIOTIC PORK

When you think of food and the U.S. military, you undoubtedly picture a long chow line where a grunt serves up chipped beef on toast, lowly privates peeling potatoes on KP duty, and semi-inedible old C-rations or more modern military field fare like palate-numbing Meals-Ready-to-Eat (MREs).

But that's the old military, not the new, modern varient—and not just because private corporations like Kellogg Brown & Root have taken over the mess halls from construction to cooking. These days, like the rest of America, the army loves to eat out. No messy preparation. No dishes to clean up. Not a chip of beef in sight. And, best of all, it's on someone else's tab. The U.S. tax-payer's. Judging by the Pentagon's own accounting, the army, navy, air force, and marines have been very hungry and they've been chowing down.

As it happens, the army has definite gastronomic tastes. Some ethnic foods, for instance, just about never make it to the table. Due to the arcane nature of the Pentagon's accounting, it is almost impossible to know for sure, but the tally on Asian food (although not Asian bases) appears to be:

Vietnamese restaurants	0
Thai restaurants	0

Indian restaurants   0
Japanese restaurants   0

And don't even ask about Afghan food!

But while it's a no-go on sushi, cooked fish is another military matter. In 2004, for instance, the army spent more than $5,000 at Chic-A-D's Cajun Chicken & Catfish Restaurant in Winnsboro, Louisiana. That same year, the catfish-hungry army dropped $6,500 at Capt'n Morgan's Steak & Catfish Restaurant in Diberville, Mississippi, and over $7,300 at Kenny's Katfish Depot in Dequincy, Louisiana. But since, as Napoleon once observed, an army marches on its stomach, the U.S. Army cannot live on catfish alone. Sandwiches are, apparently, also a must, so army eaters plunked down $13,845 at a Quiznos Classic Subs in Louisiana.

In Arkansas, the military dropped significant sums at such "Natural State" restaurants as: Rodeo Cafe ($3,485), Molly's Diner ($5,400), Annie's Family Restaurant ($8,996), and the Crispy Taco Mexican Grill ($19,283), among other establishments. While these 2004 figures were impressive, they paled in comparison to the combined sum paid out to just two El Nopal Restaurant locations in Arkansas (more than $423,000) in 2006. And for dessert, perhaps, the DoD spent a whopping $7.9 million at Arkansas's own White Dairy Ice Cream Company that same year.

But Arkansas was only a drop in the proverbial bucket (of chicken, no doubt). Military folks also sampled the fare at numerous other eateries across the country. Just a few examples from 2004:

Copper Mill Restaurant (Logan, UT)	$10,878
Bristol Bar & Grille (Louisville, KY)	$5,026
Englewood Cafe (Independence, MO)	$5,026
Pericos Mexican Restaurant (Covington, TN)	$4,050
Big Mama's Kitchen (Fayette, AL)	$3,705
Timber Lodge Steakhouse (Sioux Falls, SD)	$2,544

and some DoD favorites from 2006:

City Café (Elgin, TX)	$26,350
Home Plate Restaurant (Butner, NC)	$47,917
Pelican Café (New Orleans, LA)	$105,670

While the military clearly savors its catfish and tacos, what it really loves is barbeque! In fact, the military has sampled barbeque all across the United States—from Shotgun's Bar-B-Que Restaurant in Texas and Bo's Pit Bar-B-Que in Missouri to the Pig N' Whistle in Tennessee and Longhorn Barbecue in Washington State. In 2004, the army shelled out at least $164,828 to get its fingers greasy. In 2005 and 2006 combined, the Pentagon spent over half this amount at Corky's Bar-B-Que of Memphis, Tennessee.

While U.S. taxpayer dollars have regularly morphed into barbequed wings and ribs (with not a vegetarian restaurant in sight), the DoD wasn't completely gastronomically timid. In their travels abroad, military officials apparently did manage to sample foreign cuisine, supping at, among other places: Restaurant Schinvelderhoeve in the Netherlands ($2,133 in 2004) and Restaurante El Escudo Sociedad in Guatemala (an astounding $82,291 in 2004) and—evidently the grand champion—Singapore's First Street Café, where the DoD reportedly spent $151,883 in 2004, $216,646 in 2005, and, an astounding $310,776 in 2006, eating who knows what.

Mostly though, it's home-style comfort food and red meat in red states. In 2004, for instance, the army reportedly paid Shoney's, a purveyor of such eats as country-fried steak, chili-cheese fries, and its signature "Half-o-Pound" (a huge "chopped beef patty" adorned with "golden-fried onion rings"), more than $82,000. Just don't ask anyone to go over the top, or parachute from a plane, while that Half-o-Pound is settling.

THE *SECRET* OF COURAGEOUS CUISINE

In 2004, the Pentagon handed over $154,000 to the Secret Garden Café in Loma Linda, California. A call to the Secret Garden Café revealed that it was no longer a restaurant at all, but strictly a catering company, due to high demand from *guess who?* Its new name? Courageous Catering and Special Events. A manager at the new catering business offered the following explanation for its popularity: "We get recommended a lot because we use, like real butter, and we bring really good desserts and we only use black angus beef . . . and so they like us, and we use, like name-brand sodas instead of generics." The Army Reserve's 374th Chemical Company, she said, had just hired Courageous Catering to provide postmaneuver sustenance. For their inaugural menu, they roughed it with country-fried chicken, mashed potatoes and gravy, beans, corn on the cob with butter for dipping, fresh fruit salad, corn muffins with butter, sodas, bottled waters, iced tea, and assorted cookies and dessert bars. Hold the chipped beef, but pass the black angus and those chocolate chip cookies, Sir!

TOMMY FRANKS RIDES THE ROTISSERIE

In February 2003, *U.S. News and World Report*'s Web site reported that then four-star general Tommy Franks was said to have actually enjoyed eating MREs, but when he had his druthers he "noshe[d] at the Tex-Mex restaurant Chevys." Franks's crowning culinary moment, however, may have been in 2002 when as CENTCOM commander and the leader of U.S. forces in Afghanistan, he and Outback Steakhouse CEO Chris Sullivan decided to ship "6,700 steaks, 30,000 shrimp and 3,000 giant onions" as well as "13,400 cans of O'Douls" nonalcoholic beer, all of it donated, to members of the 101st Airborne Division, based in Kandahar. The stunt garnered Outback some meaty press coverage, so it was perhaps less than startling when Franks's decision to leave the military was quickly followed by his decision to take a spot on Outback Steakhouse's board of directors. By 2005, it was reported that he was receiving an "annual retainer of $60,000 in cash and stock," in addition to the $100,000 in restricted stock he received for joining the board.

For years, the Complex has been typified by a revolving door between the armed forces and big-time defense contractors. Franks may be pioneering a new version of this for a new military moment. Think of it as the revolving rotisserie of the military-gastronomic complex. Others may soon join him. In May 2005, just over a month after Franks officially enlisted at Outback, fellow former four-star general and chairman of the Joint Chiefs of Staff Colin Powell was the keynote speaker at the National Restaurant Association's annual conference. And the amusement park/pizzeria chain Chuck E. Cheese also seems like it's angling for a top general to join up. In fact, in 2005, it was found that all of the nearly 500 Chuck E. Cheese franchises were screening a montage of military footage, put together by the Pentagon, that even a company spokesman admitted could be interpreted as "prowar."

19

STARS AND HYPE

The *Los Angeles Times* reported, in late 2005, that the Pentagon was "secretly paying Iraqi newspapers to publish stories written by American troops in an effort to burnish the image of the U.S. mission in Iraq." It turned out that the DoD contracted "a small Washington-based firm called Lincoln Group" (which received $897,000 from the Pentagon that year and more than $697,000 in 2006), whose employees and subcontractors sometimes even impersonated reporters, to pay the newly *freed* Iraqi press to run the propaganda as legitimate news and "mask any connection with the U.S. military."

After the scandal came to light, the Pentagon launched an investigation and subsequently cleared itself of wrongdoing. According to the *New York Times,* the DoD's inspector general's report "said that the secret program . . . was lawful and that it did not constitute a 'covert action' designed to influence the internal political conditions of another country." But Tom Engelhardt, writing at the Nation.com, had his own theory on just whose opinions the Pentagon was attempting to influence. He wrote: "The goal of administration officials was always to win the war at home above all else. With that in mind, perhaps the Pentagon hired the Lincoln Group to slip those good-news pieces into the Iraqi media not to influence Iraqis, but Americans. Perhaps the hope was that the 'free' Iraqi media would be the royal route back to the American press."

Such covert ops appear to be just one front in an endless Pentagon media war. In February 2006, then secretary of defense Donald Rumsfeld launched an assault on the American media, blasting it for perceived slights against the military, such as drawing attention to the Pentagon/Lincoln Group propaganda program through an "explosion of critical press stories." The next month, he was at it again, taking the media to task for what he called a "steady stream of errors [that] all seem to be of a nature to inflame the situation and to give heart to the terrorists and to discourage those who hope for success in Iraq." Then in April, appearing on the *Rush Limbaugh Show,* he complained that top terrorists Abu Musab al-Zarqawi, Osama bin Laden, and Ayman al-Zawahiri "have media committees; they are actively out there trying to manipulate the press in the United States. They are very good at it. They're much better at managing those kinds of things than we are."

Listening to Rumsfeld's drumbeat of downcast remarks, you'd think that Al Qaeda was running the most sophisticated media empire on the planet and the U.S. military was reliant on a fourth estate that regularly acted more like a fifth column. As a result, Rumsfeld put out a call, according to the (possibly subversive) *Washington Post,* to create "24-hour media operations centers and 'multifaceted media campaigns' using the Internet, blogs and satellite television that 'will result in much less reliance on the traditional print press'" to serve the Pentagon's interests.

From Rumsfeld's remarks you'd never guess that the giants of the mainstream American media and oft-maligned news sources regularly turn up on the military's payroll. We can't be sure what the monies are for—perhaps for advertisements or just subscriptions to monitor the *enemy*—but the fact that the Pentagon is willing to fork over money to the mainstream media is at least a curiosity worthy of some reporter's attention. For example, in 2002, according to DoD documents, the military paid out thousands of dollars to the *Washington Post* and anted up tens of thousands to CNN's parent company, the Turner Broadcasting System. The next year, the *Washington Post* again received thousands of Pentagon dollars from the army, while the navy sent tens of thousands to Reuters-America. In 2004, the army paid out thousands of dollars to the New York Times Syndication Sales Corporation, while the air force paid tens of thousands in taxpayer dollars to the Public Broadcasting Service (PBS).

In 2005, the Pentagon sent tens of thousands of dollars to the Tribune Company, the parent corporation of the *Los Angeles Times* and New York's *Newsday,* among others. It also paid hundreds of thousands of dollars to ABC Radio Networks, sent the international media company Pearson more than $1.8 million, including tens of thousands earmarked for the *Economist* magazine, and paid tens of thousands to the National Newspaper Publishers Association (NNPA)—"a 65-year-old federation of more than 200 Black community newspapers from across the United States."

While some might not be surprised to learn that, in 2006, the DoD paid Rupert Murdoch's News Corporation over half a million dollars, it's also true that the Washington Post Company, the Tribune Company, Reuters, and New York's *Daily News* received money from the Pentagon as well. And in October of that year, a full-page ad showing uniformed members of the armed forces announced that the *New York Times* had partnered with the Department of Defense's Employer Support for the Guard and Reserve, along with the USO, American Legion, and others, for an upcoming job fair. Obviously, even an irritated Pentagon continues to deal with the press on a financial basis.

That may seem odd, but it gets stranger still. It turns out that the Pentagon runs a major media empire of its own. When people think about the military press, they tend to recall *Stars and Stripes* due to its World War II origins. That daily newspaper is still published today "for the U.S. military, DoD civilians, contractors, and their families." According to its own assessment, *Stars and Stripes* is "unique among the many military publications" because it is "free of control and censorship." That's an important point to remember when delving deep into the Pentagon's mammoth propaganda machine. The army, alone, publishes scores of newspapers, at posts all over the world, from Bagram, Afghanistan, to Camp Zama, Japan, including: *Advisor, Alaska Post, American Endeavor, APG [Aberdeen Proving Ground] News, Army Flier, Army Reserve Magazine, Arsenal Accents, At Ease, Azuwur, Banner, Bavarian News, Bear Facts, Belvoir Eagle, Blue Devil II, Blueridger, Buckeye Guard, Build To Serve, Bulletin, Cannoneer, Casemate, Castle Comments, Charger, Checkerboard, Citizen, Command Post, Community Report, Constellation, Corps'pondent, Countermeasure, Crosscurrents, Desert Star, Desert Voice, District Dispatch, District Observer, Dugway Dispatch, Eagle, El Morro, Engineer Update, Environmental Update, Esprit, Falcon Flier, Fir Tree, Flagship, Flaming Blade, Flash, Flightfax, Flightline/Aircraftsman, Fort Campbell Courier, Fort Dix Post, Fort Jackson Leader, Fort Leavenworth Lamp, Fort Lee Traveler, Fort Riley Post, Freedom Watch, Frontline, Georgia Guardsman, Globe, Golden Acorn, Griffon, Grizzley, Guard Detail, Guard, ME, Guard Times, Guardian, Guidon, Hawaii Army Weekly, Headwaters Highlights, Herald-Post, Herald Union, Hub, Indianhead, INSCOM Journal, Inside the Turret, Iowa Militiaman, Iraq Reconstruction Update, Iron Man, Kenny Letter, Kwajalein Hourglass, Little Rock Dispatch, McAAP News, McPherson Sentinel, MedCommunicator, Mercury, Message, Meteor-Heraut, Minuteman, Missile Ranger, Mobile, Monitor,* the *Morning Weekly, Mountaineer Defender, New York District Times, News Leader, Newscastle, Northwest Guardian, Outlook, OutPost, Pacific Connection, Pentagram, Phoenix, Plains Guardian,* the *Point, Pointer View, Prospector, Public Works Digest, Pupukahi,* the *Railsplitter, Raleigh Bulldog, Redstone Rocket, Riverside, Scout, Signal, Soundoff!, Standard, Stripe, Tarheel Minuteman, Team 19! Magazine, Tiefort Telegraph,* the *Tobyhanna Reporter, Torii, Tower Times, Triad, Tulsa District Record, Up and Down the Hill,* the *Voice, Voice of the ROK, Volkstate Guard, Warrior, Wheel,* and the *Yankee Engineer.*

The army even has its own official magazine, *Soldiers* (not to mention *Army* magazine—a publication put out by the Association of the United States Army, a "private, non-profit educational organization that supports America's Army"). The marines have their own offering, the cleverly titled *Marines* magazine, while the navy has *All Hands,* the Air Force Reserve, *Citizen Airman,* and the Defense Logistics Agency, *Dimensions*—and these are just a few of the plethora of official and quasi-official military magazines out there.

But if you think that these publications are only plying an already indoctrinated audience with so much faux news-pap, you're wrong. Officially sanctioned, thoroughly controlled, and censored military journalism forms the core of a vast public relations effort also aimed at the general public via the organs of the mainstream media. To this end, the military pushes its internal propaganda through programs like the Soldiers News Service, in which army public affairs officers "proactively engage the media to tell the human interest stories of [s]oldiers recently returned from fighting in the Global War on Terrorism." As the army readily admits, the point is to "magnify these stories by pulling them off the pages of community and installation publications and presenting them as feature story opportunities to regional and national media outlets."

In addition to its active campaign to place officially approved news pieces in civilian newspapers, the military operates a veritable online empire, with scores of propaganda-laden Web sites, like Navy.mil, the marines' USMC.mil, and GoArmy.com, each providing "news" to the public and the mainstream media, along with the Department of Defense's own site, Defenselink.mil, with its massive "press resources" section, complete with "news releases," press adviseries, news feeds, transcripts of interviews, briefings and press conferences, photos, and briefing slides. And don't forget the *objective* journalism practiced by the American Forces Press Service—a news organization operated by the Pentagon's American Forces Information Service (AFIS)—that publishes, according to AFIS, "in-depth articles . . . that explain the purpose and impact of Defense-wide programs."

The Department of Defense also clandestinely runs "news" Web sites that look like civilian operations and are not readily identifiable as the product of the U.S. military. These include Magharebia.com and the Southeast European Times, which resemble civilian news portals but, according to their fine-print disclaimers, are actually "sponsored by the US Department of Defence [*sic*]."

The DoD also has its own TV news network, the Pentagon Channel television service, which is "distributed 24/7 and is available to all stateside cable and satellite providers; via American Forces Radio and Television Service, overseas; and via webcast worldwide." An unabashed propaganda effort, the channel advertises that it broadcasts "Department of Defense news briefings, military news, interviews with top Defense officials [and] short stories about the work of our military" in order to "enhanc[e] Department of Defense communications."

The individual armed services turn out to be in on the TV act as well. For example, "Navy/Marine Corps News" is "distributed via satellite and videotape to all Navy and Marine Corps units worldwide" and, claims the navy, to "more than 320 commercial cable outlets [that] broadcast the show, giving it a huge shadow audience of family members, Reservists, retirees and civilians with an interest in the U.S. Navy." Not to be left out in the cold, the army offers "Soldiers Radio and Television."

There's also the DoD's American Forces Radio and Television Service (AFRTS). A descendant of the World War II–era Armed Forces Radio, AFRTS's mission is to "communicate Department of Defense (DoD) information to the internal audience." AFRTS's Armed Forces Network (AFN), which is run out of the military's Broadcast Center (AFN-BC), part of the Defense Media Center, at March Air Reserve Base near Riverside, California, provides seven separate television "services." These include:

AFN-Prime. Actually two networks (one devoted to the U.S. forces garrisoning Europe, the other, to those garrisoning Asia) that function like U.S. network affiliates, complete with American soap operas, game and talk shows, network sitcoms, dramas, reality shows, movies and late-night talk shows.

AFN-Spectrum. An alternative network that provides counterprogramming to AFN-Prime, including hit network shows as well as popular reruns, for the fickle military viewer.

AFN-News. Borrowing from ABC, CBS, CNN, MSNBC, NBC, and PBS, this news station spices up civilian programming with its own propaganda.

AFN-Sports. Drawing on Fox Sports channel, ESPN, FX, Speed, TBS, TNT, ABC, CBS, NBC, and Fox, this channel provides MLB, NFL, NASCAR, NHL, NCAA, and PGA sporting events.

AFN-Family. Kids' programming targeted at ages two to seventeen, offering a selection of shows from broadcast and cable networks to firmly ground junior in American culture even when living abroad.

AFN-Movie. Cleaned-up films and TV movies.

AFN-Xtra. An *extreme* network devoted to action sports, "sports events" like wrestling, talk shows, "videogamming [*sic*] programming," and "youth culture."

These networks can be viewed by military folks in 177 countries around the world.

Then there's AFRTS's radio network with content supplied by "ABC, CBS, CNN, ESPN, FOX, National Public Radio (NPR), the Associated Press Radio Network, Premier Radio Network, Westwood One, Jones Radio, Air America and USA Radio Network" and AFN's own all-news and -features channel and two sports-talk-format channels.

While the overall operation is large enough to make even right-wing media mogul Rupert Murdoch drool, it's still only a fraction of the military's vast media empire. Add to this the many publications by private organizations, veterans associations, like-minded front groups, and assorted fellow travelers catering to the military, such as the Air Force Association's *Air Force Magazine,* the National Guard Association of the United States' *National Guard Magazine,* and the Veterans of Foreign Wars' *VFW Magazine*, among many others, and you've got a propaganda machine par excellence.

But perhaps nothing drives home the immensity of the military's media holdings more strongly than the simple fact that every year, the army, alone, hands out awards for twenty distinct categories of broadcast journalism as well as twenty different types of print journalism.

In late August 2006, in the midst of Bush administration criticism of the American media for coverage of the war in Iraq, award-winning *Washington Post* national security reporter Walter Pincus (who admits to having been "'CIA-subsidized'" when he was a young freelance writer) called attention to a Pentagon request for bids for "a two-year, $20 million public relations contract that calls for extensive monitoring of U.S. and Middle Eastern media in an effort to promote more positive coverage of news from Iraq." Sounds like an effort worthy of one of the Pentagon's journalistic awards!

THE MILITARY-CORPORATE-
CONFERENCE COMPLEX

Pssst . . . ever wonder where to go to press the flesh with a genuine air force general? Or whisper sweet nothings into the ear of the chief information officer of the FBI? Or tell a key congressman on the House Armed Services Committee about the latest $2,000 cupholder retrofit for a tank? Generally, the place to be is a hotel in the greater Washington, D.C, area.

While it's rarely spoken of, there is a thriving trade in corporate-sponsored conferences *starring* personages from, or connected to, the U.S. military and intelligence communities who can be heard from—and met—for a modest price (generally under $1,500).

One major player in the military-corporate-conference-complex is the Institute for Defense & Government Advancement (IDGA), an organization that professes to be "dedicated to the promotion of innovative ideas in public service and defense." To this end, IDGA brings "together speaker panels comprised of military and government professionals while attracting delegates with decision-making power from military, government and defense industries." There, at the conferences, military big shots and their corporate cronies can cozy up to each other all weekend long.

For example, from February 28 to March 1, 2006, IDGA hosted a whole series of conferences including "Precision Strike 2006," a two-day (with an optional third, more intimate "focus day") affair

at the Hilton Washington, D.C., that first and foremost, according
to the glossy flyer mailed to those in the know, allowed attendees
to meet "military leaders and key industry players who are at the
forefront of developing precision strike capabilities, systems and
technologies." For $999 ($500 extra for the "focus day"), an enter-
prising executive could get access to: Major General Jeffrey Riemer,
commander of the Air Armament Center at Eglin Air Force Base;
Major General David Edgington, the director of global power programs for the Office of the Secretary of the Air Force for Acquisition; John S. Wilcox, the assistant deputy undersecretary of
defense for precision engagement, DoD; Colonel Daniel Johnson,
the deputy director for targets, Joint Staff; and Congressman Curt
Weldon (R-Pa.), the chairman of the Tactical Air and Land Forces
Subcommittee; among others.

Weldon showed up again, as the keynote speaker, at IDGA's
May 22–24, 2006, "Ground Combat Vehicles" conference. There,
he joined a host of armed forces brass from the army and Marine
Corps, military-academic superstars from Penn State (which
received over $135 million from the DoD in 2006) and Georgia
Tech (more than $58 million in 2006), and even Frederick W.
Kagan, a former West Point professor serving as a resident scholar
at the right-wing American Enterprise Institute.

In April 2006, IDGA hosted a three-day "Military Robotics"
conference, in cooperation with iRobot, MobileRobots (a firm that
makes robots for everyone from the army to lawn-mower king
John Deere), Northrop Grumman, and Foster-Miller, "an engineering and technology development firm" whose clients include a
long list of defense contractors and the U.S. Navy. There one might
rub shoulders with retired military personnel working in the private sector, active-duty officers from the army and navy, program
managers from DARPA, and academics from MIT and Cal Tech as
well as attend presentations on topics ranging from "Technology
Development of Unmanned Systems to Support Naval Warfighters" to "Unmanned Ground Vehicles for Armed Reconnaissance."
Closing the conference was a presentation straight out of the world
of dystopian science fiction or the *Terminator* movies: "Robots

Armed with New Technology as the Latest Weapons in Fighting the War."

While January's "Network Centric Warfare 2006" boasted such high-rolling attendees as former chairman of the Joint Chiefs of Staff General Richard Myers and former U.S. attorney general John Ashcroft, *the* event of the 2006 spring military-corporate-conference season had to be May's "GID 2006: The Geospatial Cross-Intelligence Conference for Defense" at the Westin Arlington Hotel in Arlington, Virginia. Organized by Worldwide Business Research, a concern that boasts organizing "highly focused conferences" that "facilitate informal information gathering and network creation," the three-day conference brought some of the biggest players in the military-corporate complex together with the professed goal of "advancing geospatial intelligence across all agencies to seamlessly support the warfighter."

What GID 2006 really offered, baldly in fact, was unfettered access to representatives of the CIA, FBI, Defense Intelligence Agency, army, navy, air force, and marines, among others. Specifically promised was "equal opportunity for all in attendance to gain access to DoD, federal and private sector Intelligence leaders." Sponsored by the military-corporate powerhouses Microsoft, BAE Systems (the eighth-largest defense contractor in 2006—$4.7 billion), and SGI Federal (a Pentagon favorite that provides equipment to the Warhead Performance and Target Response Branch of the Naval Surface Warfare Center), the conference included sessions hosted or cohosted by fifteen currently serving admirals or generals, as well as the Pentagon's top civilian players of that moment like Dr. Stephen Cambone, the undersecretary of defense for Intelligence, and Chuck Williams, the assistant deputy undersecretary of defense (installations), not to mention such intelligence community heavyweights as Zal Azmi, the chief information officer of the FBI, and John Kringen, the CIA's director of intelligence. The glossy conference brochure, complete with head shots of the celebrity generals in attendance, featured a picture of a military officer, drink in hand, surrounded by a bunch of guys in suits highlighting the "Welcome Cocktail Reception," where attendees would have the opportunity to "relax, unwind and network face-to-face with the delegates and speakers." The cost: just $1,498.

If testimonials on the IDGA Web site are any indication, the pay-for-play world of networking conferences is a hit with the Complex's core members. Here are but a few of the rave reviews.

> Very informative, it is great to have so many DoD and serious high level personnel to present a unified view in NCW [network-centric warfare].
>
> —EDWIN LEE, SR. PRINCIPAL ENGINEER, RAYTHEON

> Good use of time. It was very interesting hearing the various examples and lessons learned across DoD. There are a lot of common issues / lessons learned.
>
> —KEVIN FOGARTY, MANAGER, BOEING

> On target—Outstanding.
>
> —PERRY GARRETT, HEAD ENGINEER, ROCKWELL COLLINS

One attendee, who declined to offer his/her affiliation, may, however, have said it best: "Excellent. Definitely a payback on the investment."

The military-corporate-conference-complex represents just another facet of the mutually reinforcing system that undergirds the Complex. It offers a place for military brass to network with their corporate partners; to line up postservice employment contacts and paid speaking engagements after they leave their chairmanship of the Joints Chiefs or their top perch in one of the branches of the armed forces. It gives military-corporate giants a place to hold court and wannabes, of all sorts, a chance to rub shoulders with influential military figures and congressional big wigs. It's the ultimate backstage pass to the military-corporate complex and in this world, IDGA is the roadie that, for a price, will take you back to service the band.

PART VII

WEAPONIZING
THE FUTURE

In 2007, DARPA began soliciting the scientific community for proposals regarding "Chemical Robots," or "ChemBots"—soft, flexible robots that, if ever brought to fruition, are more likely to resemble Gumby than R2D2. The idea behind the project was simple. "Often the only available points of entry are small openings in buildings, walls, [and] under doors," says DARPA. In the urban war zones of tomorrow, robots made of metal or other hard materials are inadequate and need, instead, to be made "soft enough to squeeze or traverse through small openings, yet large enough to carry an operationally meaningful payload."

While ChemBots, with their expected ability to "sense and morph responsively to openings," sound like the advanced, liquid metal T-1000 model that fought Arnold Schwarzenegger's old-model Terminator in *Terminator 2: Judgment Day,* DARPA locates its inspiration in nature and the many "soft creatures, including mice, octopi, and insects" that "readily traverse openings barely larger than their largest 'hard' component." Scientists are working hard to make sure that someday the Pentagon has an army of robo-rats and cyborg moths, as well as Gumby-like ChemBots, that will crawl, fly and morph their way—alongside platoons of real-life Captain Americas—into the not-too-distant future. It's a brave new world, and the Pentagon has plenty of plans for it.

21

LIVING WEAPONS LABS

Since the invention of weapons, man has been attempting to improve them, and since World War II the United States has been the leading global actor in the research and development of weapons designed to incapacitate people. In terms of intensive research in the fields of "wound ballistics," "rapid incapacitation weaponry," and fragmentation "kill mechanisms"—what might best be called "scientific slaughter"—the United States has left the rest of the world in the dust.

Back in 1965, Jack Raymond of the *New York Times* wrote a piece aptly headlined "Vietnam Gives U.S. 'War Laboratory'"—and in that era, there were a couple of American commanders who publicly said as much. As General Maxwell Taylor, who served as the chairman of the Joint Chiefs of Staff and then U.S. ambassador to South Vietnam, put it, "We have recognized the importance of the area [Vietnam] as a laboratory. We have teams out there looking at equipment requirements of this kind of guerrilla warfare." But as Raymond pointed out, most U.S. officials were loath to make such claims for fear of comparison to the Nazis, who had, only three decades earlier, used the Spanish Civil War as a training ground for both the weapons and tactics to be used in World War II. These days, the U.S. military is even less vocal about using the planet as a living laboratory, but there's no doubt that America's recent wars, from Grenada in 1983 to Iraq in 2003, have tumbled upon each other so regularly that the military and its industrial partners have come to rely on them for battle-testing and improving their weaponry.

Military analyst William Arkin reported in the *Los Angeles Times* in 2004 that marines being deployed in Iraq were bringing along the newest high-tech gadget in America's ever-expanding arsenal to try out on whatever resistant Iraqis they happened to run into. The Long Range Acoustic Device (LRAD) emits a powerful tone that brings agonizing pain to those within earshot. While Woody Norris, chairman of the American Technology Corporation, which manufactures the device, refuses to call it a "weapon," he claims, "It will knock [some people] on their knees." But Arkin asked a crucial question seldom heard these days: "Is actual combat in a foreign country the appropriate place to test a new weapon?"

The military and its corporate partners sure think so. As the fears of the Vietnam era continue to fade, early versions of weapons are regularly rushed into battle for real-world testing, retooling, and perfecting on what increasingly seems to be a global assembly line for the military-corporate complex. For example, by 2006, press releases from American Technology Corporation noted LRAD's "capability of emitting powerful warning tones to influence behavior, gain compliance and determine intent," citing "detailed reports from Iraq" received from Marine Corps, army, and navy forces using the sonic blaster.

Other weapons follow a similar trajectory. In the mid-1990s, the Balkans became the first proving grounds for the Predator drone, an intelligence-gathering unmanned aerial vehicle (UAV). Though a contract for the Predator had only been awarded to manufacturer General Atomics (formerly a division of General Dynamics) in January 1994, a first-generation Predator was already in the skies over Bosnia by 1995. The drone then saw service in Kosovo and by 2001 had been armed with Hellfire laser-guided missiles. The Predator soon began living up to its name. In February 2001, it successfully fired one of its missiles in a flight test. Later that same year, the upgraded and armed UAV was off to the Balkans and then Afghanistan for real-world combat testing. By 2002, the Hellfire-equipped Predator drone was being used as a judge-jury-and-executioner assassination weapon in Yemen, where it attacked a civilian vehicle, incinerating six occupants—all allegedly Al Qaeda terrorists. In 2006, a Predator operating in Pakistan reportedly killed "four or five other foreign Islamic extremists" but also "13 to 18 civilians, including women and children." Today, the Predator operates in Iraq, Afghanistan, and elsewhere. Says Major Russell Lee of the air force, "There is always a Predator airborne around the world."

In 2007, another drone entered the proving grounds: the MQ-9 Reaper. These jet fighter–sized, turboprop "hunter-killer" UAVs, each capable of carrying one and a half tons of bombs and missiles, were heralded by the Associated Press as part of "aviation history's first robot attack squadron" whose deployment would "be a watershed moment even in an Iraq that has seen too many innovative ways to hunt and kill." In actuality, Reapers were reportedly slated for battle-testing in Afghanistan first, before being sent on to occupied Iraq. In both war zones, their deployment will mark a leap forward in lethality as, in addition to being able to fly twice as high and fast as the Predator, Reapers can carry some fourteen Hellfire missiles compared to the older drone's maximum of two. Not surprisingly between May and August 2007, alone, the Pentagon awarded over $291 million in Reaper-related contracts to General Atomics (which *reaped* over $669 million from the DoD the previous year).

While the Predator has already seen plenty of service in Iraq and the Reaper is reportedly on its way, the so-called Mother of all Bombs, the 21,500-pound Massive Ordnance Air Blast superbomb (MOAB), arrived just barely too late for the March 2003 shock-and-awe assault on that country, despite a very public rush to ready it for the war. As the BBC noted, the weapon, which was 40 percent more powerful than the next largest conventional bomb in the

U.S. arsenal, "spreads a flammable mist over the target and then ignites it, creating a highly destructive blast, which experts say packs the force of a small nuclear bomb."

Unfortunately only for MOAB's makers at the air force Research Laboratory, the advance on Baghdad happened too swiftly. Although a single bomb was readied for use in April 2003 (after tests at Florida's Eglin Air Force Base) and sent to an "undisclosed forward base" in the Iraq "theater," it didn't reach the region in time to vaporize anyone or anything. Since then, however, it has reportedly been sitting in the "Iraq war region," presumably waiting to be unleashed on the next evildoer nation or regional rogue state.

In 2007, however, even the MOAB was bested by a new Boeing product—the Massive Ordnance Penetrator, or MOP. In October of that year, some $300 million, part of a $42.3-billion presidential appeal to fund for the wars in Iraq and Afghanistan, was specifically requested to fund technologies including this 30,000-pound mega–bunker buster that can, reportedly, blast through two hundred feet of concrete.

During the Vietnam War, the U.S. military tried out all sorts of half-baked high-tech weapons systems, such as Robert McNamara's famed "electronic battlefield" of remote sensors and land mines and the various "people-sniffing" devices (from live bedbugs to chemical-mechanical apparatuses). These were tested in the field with less than stellar results. But there were notable successes as well such as the M-16 rifle and a new generation of enhanced antipersonnel munitions (like advanced cluster bombs and napalm-B, the jellied gasoline that burned hotter and longer than its predecessor). Basically, whatever could be tried out in action was.

Still, the military was forced, by the rise of an antiwar movement, to engage in a debate, of sorts, about the weapons it was using and felt the weight of negative public opinion when employing chemical gases, napalm, and defoliants. Today, the military is remarkably unintimidated, and there's almost no debate about using the United States' seemingly endless string of wars (including the possibly never-to-be-ended Global War on Terror) as so many proving grounds for whatever UAVs, superbombs, or sonic-scream devices the Complex can come up with. Thanks to nearly nonstop conflicts, interventions, engagements, and attacks during the presidential administrations of both parties, what was once taboo is now the norm. In fact, Gerry J. Gilmore of the American Forces Press Service, in discussing the statements of a senior Defense Department official, used a stock phrase from the Vietnam War to make the case for employing new LRAD-type technologies. These were, he said, meant "to win hearts and minds during 21st-century military operations." No uproar followed.

It is worth remembering that some of the same chemical gases unleashed in Vietnam were also used, at home, on demonstrators at street protests and on college campuses. The same may well turn out to be true of today's technologies. The New York Police Department was ready with the Long Range Accoustic Device for the 2004 Republican National Convention protests in the Big Apple; in 2005, it was sent to "areas hit by Hurricane Katrina" for "crowd control" purposes; and by 2006 LRAD was in the hands of U.S. Border Patrol agents. The only question now is: When will its eardrum-shattering tones be brought to bear on civilians in the U.S. "homeland"?

22

THE WILD WEAPONS
OF DARPA

When, in October 1957, the USSR launched the first man-made satellite, the basketball-sized *Sputnik,* it caught the United States off guard and sent the government into fits. Not only had the Soviets exploded an atomic bomb years before the United States predicted they would, but now they were leading the "space race." In response, the Defense Department approved funding for a new U.S. satellite project, headed by the former Nazi rocket scientist and SS officer Wernher von Braun, and created, in 1958, the Defense Advanced Research Projects Agency (DARPA) to make certain that the United States forever after maintained "a lead in applying state-of-the-art technology for military capabilities and to prevent technological surprise from her adversaries."

Half a century later, the United States stands alone as the globe's sole hyperpower. Yet DARPA, the agency for an arms-race world, seems only to be warming up to the chase. There may be no country left to take the lead from the United States, the nearest military competitor being China, which reported $45 billion in defense spending in 2007 (or $85–125 billion according to Pentagon estimates), while real U.S. national security spending approaches $1 trillion per year. In 2003, when China put its first "Taikonaut" into outer space, the United States was spending about as much on weapons development as any other nation was spending on its

entire military budget. Undaunted, DARPA continues to develop
high-tech weapons systems for 2025–2050 and beyond—some of
them standard fare like your run-of-the-mill hypersonic bomber,
others more exotic.

In 2003, the *Los Angeles Times* reporter Charles Piller noted that
DARPA had put forth some of the "most boneheaded ideas ever to
spring from the government"—including a "mechanical elephant"
that never made it into the jungles of Vietnam and telepathy
research that never quite afforded the United States the ability
to engage in psychic spying. As former DARPA director Charles
Herzfeld noted in 1975, "When we fail, we fail big." Little has
changed. According to DARPA's current chief, Anthony J. Tether
(who formerly worked for major defense contractors like Science
Applications International Corporation and Loral Corporation),
some 85 to 90 percent of its projects fail to meet their full objectives. Still, Piller pointed out, DARPA "has been behind some of
the world's most revolutionary inventions," including "the Internet, the global positioning system, stealth technology and the
computer mouse."

DARPA's spectacular failure rate and noteworthy successes stem
from its high-risk ventures. For years, DARPA has funded extremely
unconventional, sometimes beyond-the-pale, research in all realms
of science and technology. It is, perhaps, the most creative place in
the U.S. government for a scientist who wants to stretch his or her
mind in adventurous directions and be well paid to do so. If you
have a wild idea, DARPA's the place to try it out. Said the Harvard
University pathologist Donald Ingber in 2001, "DARPA [has] funded
things that a lot of people thought were ridiculous, and some that
people thought were impossible. They make things happen."

There's only one caveat: In one way or another, just about
every project, however mind stretching, invariably must aid,
directly or indirectly, in the injury or death of people the world
over. Consequently the projects are often some of the most lethal
ever conceived. Over the years, DARPA research has led to a
plethora of products designed to kill, among them the M-16 rifle,
Hellfire-missile-equipped Predator drones, stealth fighters and

bombers, surface-to-surface artillery rocket systems, Tomahawk cruise missiles, B-52 bomber upgrades, Titan missiles, Javelin portable "fire and forget" guided missiles, and cannon-launched Copperhead guided projectiles.

A question seldom asked is why unfettered, blue-skies creativity is fostered only in the context of lethal technologies (or those that, indirectly, enhance lethality by aiding the functioning of the armed forces). As the United States continues its mad dash into a post–Cold War, one-nation arms race, the answer can't be fears of a missile gap or the menace of a technologically advanced foreign foe. Nor can it be a general rush to keep skills from falling behind the rest of the world. Just look at the state of education in America. On UNICEF's 2002 list of teens in industrialized countries falling below international academic benchmarks, the United States ranked seventh worst out of twenty-four nations. In 2003, U.S. teens tested in math scored lower than twenty of twenty-nine industrialized nations. Despite the poor showing, no one rushed to set up an Advanced Education Research Agency.

The United States is the "largest single emitter of carbon dioxide from the burning of fossil fuels," according to the CIA's annually published *World Factbook,* yet the Environmental Protection Agency's National Center for Environmental Innovation is a far cry from a DARPA-like entity. In 2007, for example, it doled out a mere $1.62 million in seven state-innovation grants in 2007. DARPA, by comparison, spends about $3 billion each year on some two hundred projects that range from human performance enhancement to unmanned aerial vehicles. But just because the government isn't pouring money into the projects of scientists eager to attack environmental problems doesn't mean its not interested in environmental research. In fact, DARPA has taken up the torch and is funding a rigorous research program aimed at finding novel ways to weaponize the natural world.

As evidenced by its Vietnam-era mechanical elephant project and a more recent grant to researchers developing a robotic canine called Big Dog for the army, DARPA might be said to have something of an animal fetish. Recent ventures whose very names evoke the ethos of the wild kingdom include:

Cormorant. An unmanned aerial vehicle (UAV) program to "examine the feasibility of a UAV that may be deployed from the sea without carrier support."

Falcon. A program to "develop and demonstrate hypersonic technologies that will enable prompt global reach missions" via a "reusable Hypersonic Cruise Vehicle (HCV) capable of delivering 12,000 pounds of payload a distance of 9,000 nautical miles from CONUS [the continental United States] in less than two hours."

Hummingbird Warrior. A program to produce a helicopter-like vertical takeoff and landing UAV.

Piranha. A project to "enable submarines to engage elusive maneuvering land and sea targets."

Panda (Predictive Analysis for Naval Deployment Activities). A project that seeks to locate maritime vessels "deviating from their normal, expected behavior in ways that may be indicative of an emerging threat."

Walrus. A program to develop a very large airlift vehicle, capable of hauling a payload capacity of approximately five hundred tons over global distances.

Wasp. A hand or "bungee-launched" micro air vehicle of less than two hundred grams with a twelve-inch wingspan.

WolfPack. A group of miniaturized, unattended ground sensors that are meant to work together in detecting, identifying, and jamming enemy communications.

The agency also embraces the imagery of the natural environment in its Organic Air Vehicles in the Trees project, which sounds downright *green*, although it's actually meant to create tiny UAVs that will fly through forests, over hills, and around cities searching for enemies. There's also the ecofriendly-sounding FORESTER, a program "to design, fabricate, integrate and test a system that can detect and track moving dismounted soldiers (as well as vehicles) under foliage to a range of at least 30 km."

Allusions to the natural world, however, are the least of it. While the military is well versed in employing all sorts of creatures to do its bidding from army guard dogs to navy dolphins trained to locate sea mines, DARPA is keen on branching out from or *improving* on the class Mammalia. One way is through its Bio-Revolution Program, which seeks to "harness the insights and power of biology to make U.S. warfighters and their equipment . . . more effective."

WILLARD AND HIS WILD PALS

Killer Bees

After all those years of warnings about sinister African killer bees inexorably heading toward the United States, in 2002 DARPA decided to draft bees into military service, launching projects to examine the performance of honeybees trained to detect explosives and locate other "odors of interest." Since then, DARPA has been creating insect databases while increasing efforts to "understand how to use endemic insects as collectors of environmental information." DARPA says it has already tested "this endemic insect system in key operational demonstrations here and abroad." How long until it starts thinking about weaponizing insects as well? Instead of your plain old, garden-variety Stinger missiles, you could have a swarm of missile stingers.

Giving You the Fish Eye

DARPA's Bio-Optic Synthetic Systems Program seeks "to demonstrate new bio-inspired concepts in optics" for use in "military optical systems such as UAVs or missile guidance components." One of the program's inspirations: "crystalline fisheye lenses."

(Octo)Pie in the Sky Camouflage

According to the agency's 2003 strategic plan, "DARPA-supported researchers are studying how geckos climb walls and how an octopus hides to find new approaches to locomotion and highly adaptive camouflage. The idea is to let nature be a guide toward better engineering." Imagine the ink-squirting, suction-cup-covered frogman of the future!

Remote-Control Roborats and Mechanical Moths

In 2002, DARPA researchers demonstrated that, using a laptop computer, they could remotely control the movements of a rat with electrodes implanted in its brain. In 2003 and 2004, DARPA's Robolife Program researchers turned their attention to the "performance of rats, birds and insects in performing missions of interest to DoD, such as exploration of caves or covert deposition of sensors." By early 2006, DARPA had put out a call for "proposals to develop technology to create insect-cyborgs, possibly enabled by intimately integrating microsystems within insects, during their early stages of metamorphoses." According to the solicitation, while DARPA was particularly interested in "flying insects," such as moths and dragonflies, it would also accept proposals focusing on "hopping and swimming insects" implanted with microphones and video sensors. The next year, news broke that DARPA researchers were "raising cyborg beetles."

A 2007 article in the *Times* (UK) revealed that moths would, in the not too distant future, be implanted with computer chips while still in their cocoons, allowing their nervous systems to be remotely controlled. These cyborg moths (Micro-Electro-Mechanical Systems, or MEMS in DARPA-speak) could then be flown into homes, bases, or anywhere else the Pentagon might want to spy and beam back video and other data. "This is going to happen," said Rodney Brooks, director of the computer science and artificial intelligence lab at MIT, which is working on the project. In fact, it might even have already happened. Later that year, the *Washington*

Post reported on accounts, stretching back to 2004, of "dragonflies" spotted at political protests that some believe are "insect-size spy drones" deployed by the Department of Homeland Security or other government agencies.

Even if the robobugs haven't been deployed yet, they're certainly on the way. Brooks, also the cofounder and chief technology officer of defense contractor iRobot, notes, "It's not science like developing the nuclear bomb, which costs billions of dollars. It can be done relatively cheaply." The insects might even be weaponized—imagine a swarm of cyborg suicide moths. Brooks went on to note, "The DoD has said it wants one third of all missions to be unmanned by 2015, and there's no doubt their things will become weaponised, so the question comes: should they [be] given targeting authority?" He continued, "Perhaps it's time to consider updating treaties like the Geneva Convention to include clauses which regulate their use."

Militarizing the animal world, however, carries its own risks. Take World War II's Project X-Ray, in which bats with incendiary explosives strapped to their bodies turned on their military masters and set fire to a U.S. Army airfield. Just imagine what an army of infantry insects or army rats might do! Anybody remember the movies *Them!* or *Willard*?

Shark-bot

In 2006, *New Scientist* magazine revealed that DARPA researchers had perfected "a neural implant designed to enable a shark's brain signals to be manipulated remotely, controlling the animal's movements" by which "they hope to transform the animals into stealth spies, perhaps capable of following vessels without being spotted." Their plan was to "implant the device into blue sharks and release them into the ocean off the coast of Florida." Call Police Chief Brody back to duty!

Fee, Fi, Faux Fauna

DARPA's "Stealthy Sensors Program" "seeks to exploit the revolutionary sensing and mobility of animal sentinels for unique defense applications" and "the natural ability of animal systems in training and learning to track, detect, and deliver to targets."

Little Shop of Horrors

There seems to be no end to DARPA's inventiveness. Through the Biological Sensory Structure Emulation (BioSenSE) Program, researchers are attempting to create synthetic versions of nature's "sensory structures" that detect stimuli like changes in temperature and pressure. The "majority of these stimuli are of great military relevance," they claim. DARPA has also held a workshop to "help researchers in various disciplines self-assemble into teams capable of developing plant inspired actuation systems that will ultimately have application in military adaptive or morphing structures." What's on the horizon then? A brigade of Swamp-Thing warriors?

Finally, there's also a new *cold war* being planned out in DARPA's Special Projects Office through its Polymer Snow project—a program to create a polymer-based synthetic "snow," with "the look and feel of natural snow," that is "tailored to reversibly control vehicle and personnel mobility." This faux snow will turn whatever areas the military chooses into a winter wonderland where movement is impeded and people are left snowed in.

THE WILDEST OF APES

Perhaps the most frightening of DARPA's weaponized science projects are those that deal with militarily enhancing that most violent of apes—man. In 2003, for instance, DARPA touted the "Enhanced Human Performance" component of its Bio-Revolution Program, whose aim was to prevent humans from "becoming the weakest link in the U.S. military." Lest rats, bees, and trees become the dominant warriors, Enhanced Human Performance proclaims it will "exploit the life sciences to make the individual warfighter stronger, more alert, more endurant, and better able to heal." Yes, what now captivates DARPA researchers once captivated comic-book readers: the dream of creating a real-life Captain America, that weakling-turned-Axis-smashing-superpatriot by way of "super-soldier serum."

Just Say "No" to No Doze, but "Yes" to Endless Combat

The U.S. military has long plied its fighting men with uppers. In Vietnam, medics satisfied soldiers' need for speed by doling out government-issue amphetamines. In 2002, U.S. pilots under the influence of air force "go-pills" (which the air force spokeswoman Lieutenant Jennifer Ferrau called a "fatigue management tool") killed four Canadian soldiers and injured eight others when they dropped a laser-guided bomb on a Canadian military training exercise in Afghanistan. Today, DARPA's Continuous Assisted Performance (CAP) Program is aimed at creating a 24-7 trooper by "investigating ways to prevent fatigue and enable soldiers to stay awake, alert, and effective for up to seven days straight without suffering any deleterious mental or physical effects and without using any of the current generation of stimulants."

This is Your Brain on DARPA . . . Any Questions?

DARPA researchers are also at work on a Brain Machine Interface ("neuromics") project, designed to allow mechanical devices to be controlled via thought-power. Thus far, researchers have taught monkeys to move a computer mouse and a telerobotic arm simply by thinking about it. With up to ninety-six electrodes implanted in their brains, the monkeys are able to reach for food with a robotic arm. Researchers have even transmitted the signals over the Internet, allowing remote control of a robotic arm six hundred miles away. In the future they hope to develop a "non-invasive interface" for human use. Says DARPA: "The long-term Defense implications of finding ways to turn thoughts into acts, if it can be developed, are enormous: imagine U.S. warfighters that only need use the power of their thoughts to do things at great distances." For years, the U.S. military has been improving its ability to reach out and kill someone. Is its mantra for the future: *If you think it, they will die?*

LIFE (AND DEATH) SCIENCES

Speaking about a "bio-optics" project to enhance weaponry, Leonard J. Buckley, a program manager in materials chemistry at DARPA's Defense Sciences Office, said, "Inspiration from nature . . . will allow more life-like qualities in the system." DARPA spokeswoman Jan Walker elaborated more generally: "We're inter- ested in investigating biological organisms because they have evolved over many, many years to be particularly good at surviv- ing in the environment . . . and we hope to learn from some of those strategies that Mother Nature has developed."

Poor Mother Nature. What hope has she when faced with yearly real defense spending edging toward $1 trillion? What can she do when the most powerful thing driving freethinking U.S. sci- entists is the urge to weaponize her offspring? Under DARPA, the life sciences have become a fertile area for furthering the science of slaughter in an effort, in the words of the DARPA Defense Sciences Office, to overcome the "Frailties of Life" to achieve "Super Physio- logical Performance." How superbly Nietzschean!

Such is the state of government-sponsored innovation in the United States. If you're a researcher in crucial fields and want the time, funding, and latitude to be creative, your work must benefit the Pentagon in its race to make sure that the next Saddam Hussein can be, in the words of Major General Raymond Odierno, "caught like a rat" by one of the army's robo-rat patrols.

Other than finding new ways of circumventing international law (bypassing violations of national airspace, for instance, with space-launched weapons), which the United States already does quite well with current technology, it's hard to fathom why the government is still locked in a Cold War–style arms race in a single hyperpower world. The only possible explanation—other than the mountain climber's mantra "because its there"—lies in the driving will of the ever-expanding military-corporate complex. This would certainly help explain why, for instance, the Environmental Protection Agency's entire 2006 research and development budget was $622 *million* while the DoD's was $73.7 *billion*.

For today's researchers, DARPA is, both intellectually and financially, a fabulous and alluring gravy train. The freedom to dream and create, DARPA's mandate, is seductive and, as such, so dangerous that it's hard not to suspect that war making is now America's most advanced product.

23

CAPTAIN DARPAMERICA

Even if you never read the comic book or watched the hopelessly low-production-value 1960s cartoon, chances are you've at least seen the image of Captain America—the slightly ridiculous-looking superhero in a form-fitting, star-spangled bodysuit who was killed off by Marvel Comics in 2007. If you're still hazy on "Cap," he was Steve Rogers, a 4-F weakling during World War II who, through the miracles of modern science (a "super soldier serum"), became an Axis-smashing powerhouse—the pinnacle of human physical perfection and the ultimate American fighting man.

In the 1940s comic, Rogers had taken part in a supersoldier experiment, thanks to the interventions of an army general and a scientist in a secret government laboratory. He was to be the first of many American supersoldiers, but due to poor note-keeping methods and the efforts of a Nazi assassin, he became the sole recipient of the serum. Today, however, the dream of Captain America turns out to be alive and well—and lodged in the Pentagon. The U.S. military aims to succeed where those in the four-color comic book world failed. By using high technology and cutting-edge biomedicine, the military hopes to create an entire army of Captain Americas—a fighting force devoid of "Steve Rogers" or even "Average Joes" and made up instead of supersoldiers whose messy, laggard humanness has been all but banished.

TWENTY-FOUR-HOUR SOLDIERS

The military has long been interested in creating an always-on, twenty-four-hour fighting man. During the Vietnam War, the army undertook extensive studies on the effects of sleep deprivation. At the time, however, all the military could offer was copious amounts of amphetamines to keep men wired for combat. Today, with the military again stretched thin and a manpower crisis looming on the horizon, the quest for a two-for-the-price-of-one soldier who never needs to sleep is even more urgent. To this end, DARPA started a "Preventing Sleep Deprivation Program." Its aim was to work on ways to enable a pilot "to fly continuously for 30 hours," a Green Beret to carry out forty-eight to seventy-two hours of sustained activity, or "advancing ground troops [to] engage in weeks of combat operations with only 3 hours of sleep per night"—all without suffering from cognitive or psychomotor impairments.

Scientists in the Complex are hard at work on this line of research. At Wake Forest University (which received more than $760,000 in DoD funding in 2006), for instance, researchers are studying a class of medicines known as Ampakines, which are thought to protect against the cognitive deficits ordinarily associated with sleep deprivation. At Columbia University (over $1.2 million from the DoD in 2006), new imaging technologies are being employed as part of a program to study the "neuro-protective and neuro-regenerative effects" of an antioxidant found in cocoa. (In low-tech World War II, the military just gave the grunts chocolate bars.) Who's conducting this line of research for DARPA? Why, researchers at the Salk Institute and also at that all-chocolate-all-the-time company Mars—yes, the folks who bring you M&M's and Snickers! (In 2005–2006, Mars received well over $100 million from the Pentagon.)

At the same time, the Air Force Research Laboratory's Warfighter Fatigue Countermeasure Program is looking into a drug called Modafinil, which can reportedly keep people awake for up to eighty-eight hours; while researchers at the Naval Health Research Center (NHRC), the Space and Naval Warfare Systems Command (SPAWAR), the Walter Reed Army Institute of Research, and the U.S. Army Aeromedical Research Laboratory, among others, are working on sleep- (or-lack-thereof)-related projects. This despite the fact that in 2007, another team from Walter Reed published a study demonstrating that a lack of sleep clouds moral judgment.

MAJOR MORALITY, YOU'RE DEMOTED. WE'RE PROMOTING CORPORAL PUNISHMENT!

Sleepless soldiers are all well and good while the fighting goes on; but how does one prevent sleepless, anxiety-filled nights after those missions end? Once upon a time, it seems, most soldiers felt a great revulsion at close-quarters killing. During World War II, it has been estimated, as few as 15 to 20 percent of American infantry troops actually fired their weapons at the enemy. By the Vietnam years, the military had managed to bring that number up into the 90 to 95 percent range. Obviously, the armed forces had found ways to turn American men into more efficient killers. But how to deal with the pesky problems of regret, remorse, and post-traumatic stress disorder?

Journalist Erik Baard raised the specter of the creation of a "guilt-free soldier," noting that researchers from various universities across the United States—including Harvard (over $5.1 million from the DoD in 2006), Columbia, New York University (more than $969,000 in 2006), and the University of California at Irvine—were working on various methods of fear inhibition and also memory numbing by using "propranolol pills . . . as a means to nip the effects of trauma in the bud." He further reported that, at Columbia, the lab of Nobel laureate in medicine Eric Kandel had "discovered the gene behind a fear-inhibiting protein, uncovering a vision of 'fight or flight' at the molecular level." When asked by Baard if he was funded by DARPA, Kandel answered, "No, but you're welcome to call them and tell them about me."

REMOTE-CONTROLLED SOLDIERS?

As noted in a 2004 *New Yorker* article, the military, searching for perks to retain troops, began offering free cosmetic surgery (funded by taxpayer dollars) to anyone "wearing a uniform." So right now "bigger breasts" are the type of implants the U.S. military is specializing in. (Military doctors performed 496 breast enlargements between 2000 and 2003.) However, if DARPA scientists have their way, the implants du jour of the future may be the product of the Brain Machine Interface Program, and its "new high-density interconnects for brain machine interfaces that will allow [researchers] to monitor the brain patterns associated with a wide variety of behaviors and activities relevant to DoD."

DARPA already has all sorts of programs designed to *improve* humans. Take the Neovision Program whose goal is "using synthetic materials for a retinal prosthesis to enable signal transduction at the nerve/retina interface"; that is, creating devices to technologically enhance or even reconceptualize human vision as we know it. Or how about the Biologically Inspired Multifunctional Dynamic Robotics (BIODYNOTICS) Program, which aims to develop "robotic capabilities," inspired by biology—such as the movements of arms and legs—"for national security applications."

FOODLESS FIGHTERS? WATER-FREE WARRIORS?

But what good is an always-on, morals-free cyborg soldier if s/he's caught in the classic quagmire of having recurring desires to eat and drink, which simply must be met? How pathetically human! Not to worry. Today's soldiers might complain about choking down MREs, but if all goes well, tomorrow's troops won't worry about missing a meal.

Typical adults require about 1,500 to 2,000 calories per day, but Special Forces troops may require as many as 6,000 to 8,000 calories per day while in the field. Taking time to eat, however, wastes time, so DARPA's Peak Soldier Performance Program is investigating ways of "optimizing metabolic performance" to achieve "metabolic dominance" and allow future soldiers to operate at "continuous peak physical performance and cognitive function for 3 to 5 days, 24 hours per day, without the need for calories."

At the same time, the DARPA crew has instituted the Water Harvesting Program, which seeks to "eliminate at least 50 percent of the minimum daily water supply requirement (7qts/day) of the Special Forces, Marine Expeditionary Units, and Army Medium-Weight Brigades" through initiatives such as deriving "water from air." Perhaps someday soldiers will be able forgo water altogether for long periods of time thanks to the efforts of the Combat Feeding Directorate of the U.S. Army Soldier Systems Center in Natick, Massachusetts. Yes, the lab that created the "indestructible sandwich" (which boasts a three-year shelf life) has since come up with a dried-food ration that troops can hydrate by urinating on it. And you thought military food was piss-poor to begin with?

SUPERSUITS: CAN I GET THIS IN STAR-SPANGLED SPANDEX?

What can you say about Captain America's outfit? While certainly distinctive, his red, white, and blue threads were always a bit light on function. But what of the Captain Americas of the future? They won't be clad in jingoistic jumpsuits. The army's Natick Soldier Systems Center is currently supervising a seven-year, $250-million Future Force Warrior Program, set to be rolled out in 2010, which will outfit soldiers with new, lighter body armor, an on-board computer, "e-textile" clothing (with wiring for computer systems woven into it), a helmet with built-in night vision, a computer screen monocle, and bone-conduction microphones. Fast forward a decade to the military's Vision 2020 Future Warrior system, and you'll find an all-black, sci-fi, storm-trooper outfit that looks like it came from a B-movie prop trailer. Sadly, both 2010 and 2020 are already looking so last year, before they've even had a chance to encase a military body.

Future Warrior Concept (2020). *Photo by Sarah Underhill. Courtesy of the U.S. Army Natick Soldier Systems Center.*

In 2004, Steven G. Wax, the director of DARPA's Defense Sciences Office, addressed members of the academic, corporate, and military communities and told them that the mech-suit worn by Sigourney Weaver in the movie *Alien* was fast becoming a reality. While various clunky exoskeletons have been produced since the 1960s, Wax indicated that "breakthroughs in structures, actuators and power generation—with a bit of help from advanced microelectronics"—have now made it possible to create a workable "external structure that can move unobtrusively with a soldier and still carry more than 100 pounds with no effort by the wearer." Through its Exoskeletons for Human Performance Augmentation Program, DARPA claims to be en route to creating even more advanced "self-powered, controlled, and wearable exoskeleton devices and/or machines" specifically designed, of course, to "increase the lethality" of U.S. soldiers.

FOOD FOR THOUGHT

There's no DARPA-esque organization involved in actually solving the most pressing global problems. In a world where many still lack access to adequate clothing, DARPA is pouring massive sums into building costly robotic suits. In a world where 800 million people suffer from malnutrition and 1 billion lack access to potable water, DARPA is spending vast amounts of money to figure out how a few (well-armed) people in the global North can do without food or water on military missions (generally in the global South). And yes, while some in the developing world could benefit from possible DARPA spin-off and trickle-down innovations like futuristic prosthetic limbs, many, many more could benefit from low-cost, low-tech public health initiatives. Of course, many would have no need for high-tech prosthetics if, for so many years, the U.S. military hadn't pumped so much money into weaponry development, especially land-mine research and production. (For instance, as many as 3 million land mines and "800,000 tons of war-era ordnance" may still lie in Vietnam, more than thirty years after the U.S. intervention there ended. Civilians continue to be maimed or killed as a result.)

DARPA's chunk of the vast Pentagon budget is a cool $3 billion, a sizable hunk of which is now being devoted to creating real-life Captain DARPAmericas. Like so many of the agency's projects, its efforts to craft the supersoldiers of tomorrow typify the ultimate in sci-fi thinking.

In reality, however, most DARPA projects fall far short of meeting their ultimate goals. During the Vietnam War, massive amounts of money, firepower, and high-tech weaponry proved unable to stamp out an enemy that regularly used punji sticks (sharpened bamboo) as a weapon. In Iraq, billions upon billions of dollars in military and intelligence spending for satellites, state-of-the-art surveillance devices, stealth bombers, fighter jets, UAVs, tanks, Bradley Fighting Vehicles, Humvees, heavy weapons, night-vision devices, high-tech drones, experimental weaponry, and all the trappings of the Complex's signature brand of Technowar, though capable of killing large numbers of people, have not stopped insurgents who lack heavy armor, airpower, spy satellites, body armor, and high-tech gear and fight with AK-47s—a rifle designed in the 1940s—pickup trucks, and bombs detonated by garage-door openers.

As a key player in the Complex, DARPA imagines the future through the lens of the present. Strangely enough, its projects, however scientifically visionary, are hampered, at their core, by the very opposite of blue-sky thinking—an inability to get beyond the military mind-set and premises of today (or even yesterday). Where Pentagon seers envision an army of unstoppable comic-book heroes, they may well find overwrought, strung-out soldiers, suffering from the still-unknown side effects that are sure to come from interfering with basic human functions like sleeping and eating. The superhero they've togged out in high-tech gear may well prove vulnerable to as yet undeveloped, but sure to be cheap, crude, and effective jamming devices and countermeasures. Odds are, the Pentagon would be better off investing in actual Captain America outfits. Not only would it be infinitely cheaper, but who's gonna mess with a platoon clad in star-spangled spandex?

24

BAGHDAD, 2025

So you think that American troops, fighting in the urban maze of Baghdad's huge Shiite slum, Sadr City, reflect nothing more than a horrible strategic mistake, an unexpected fiasco? The Pentagon begs to differ. For years now, its planners have believed that guerrilla warfare, not against Guevarist *focos* in the countryside of some recalcitrant land but in the urban "jungles" of the vast slum cities that increasingly cover the planet, is where the future lies. Take this description of combat in an urban labyrinth:

> Indigenous forces deploying mortars transported by local vehicles and ready to rapidly deploy, shoot, and re-cover are common . . . [Meanwhile,] an infantry company as part of the US rapid reaction forces has been tasked with the . . . mission to secure several objectives including the command and control cell within a 100 square block urban area of the capital.

Is it Baghdad? It's certainly possible, since the passage was written in 2004 with urban warfare in Iraq's capital already an increasingly grim reality for Washington's military planners. But the actual report—by a DARPA official—focused on cities-of-the-future, of the year 2025 to be exact, as part of "a new DARPA thrust into Urban Combat."

Fear of urban warfare has long been an aspect of American military planning. Planners remember urban killing zones of the past where U.S. forces sometimes suffered grievous casualties. These included Hue, South Vietnam's old imperial capital, where "devastating" losses were incurred by the marines in 1968; the Black Hawk down debacle in Mogadishu, Somalia in 1993, where local militias inflicted 60 percent casualties on army Rangers; and, of course, the still-ongoing catastrophe in Iraq's cities. Prior to the Bush administration's 2003 invasion of Iraq, American newspapers were full of largely military-leaked or inspired fears that, as Rajiv Chandrasekaran wrote in the *Washington Post* in late September 2002, Saddam Hussein "would respond to a U.S. invasion by attempting to . . . draw U.S. forces into high-risk urban warfare." It was feared that the taking of "fortress Baghdad," as then defense secretary Donald Rumsfeld termed it, might prove costly indeed.

On April 8, 2003, however, the *Washington Post* reported that "U.S. Army troops rolled into Baghdad," and conventional wisdom, in and out of the administration, held that "victory"—the very name given to the first major base the United States established in Iraq, right at the edge of Baghdad International Airport—was close at hand. That was then, of course. On October 8, 2006 (three years and six months later, to the day), the *Post* confirmed that the worst preinvasion fears of military planners had, in fact, come true—even if somewhat belatedly and with Saddam Hussein imprisoned somewhere in the confines of Camp Victory.

The "number of U.S troops wounded in Iraq," wrote reporter Ann Scott Tyson, "has surged to its highest monthly level in nearly two years as American GIs fight block-by-block in Baghdad." In fact, aside from the Sunni stronghold of Anbar Province, Baghdad had, by then, become the deadliest location for U.S. troops in Iraq, and urban warfare in a slum city, involving snipers, IEDs, suicide car bombs, and ambushes of all sorts had, it seemed, become America's military fate. Midway through the next year, Tyson would note that, as more U.S. troops *surged* into Baghdad, Iraqi urban guerrillas began employing "increasingly sophisticated and lethal means of attack . . . resulting in greater numbers of American fatalities relative to the number of wounded"—making May 2007 the third-deadliest month for U.S. troops since the invasion.

DARPA'S FUTURE WAR ON THE URBAN POOR

In *Planet of Slums,* Mike Davis observes: "The Pentagon's best minds have dared to venture where most United Nations, World Bank or State Department types fear to go . . . [T]hey now assert that the 'feral, failed cities' of the Third World—especially their slum outskirts—will be the distinctive battlespace of the twenty-first century." Pentagon war-fighting doctrine, he notes, "is being reshaped accordingly to support a low-intensity world war of un-limited duration against criminalized segments of the urban poor."

In October 2006, the army issued an updated "urban opera-tions" manual. "Given the global population trends and the likely strategies and tactics of future threats," it declared, "Army forces will likely conduct operations in, around, and over urban areas— not as a matter of fate, but as a deliberate choice linked to national security objectives and strategy, and at a time, place, and method of the commander's choosing." Global economic deprivation and poor housing, the hallmarks of the urban slum, are, the manual asserted, what makes "urban areas potential sources of unrest" and thus increases "the likelihood of the Army's involvement in stabil-ity operations." The manual's authors were particularly concerned about "idle" urban youth loosed, in the future slum city, from the "traditional social controls" of "village elders and clan leaders," and, thus, prey to manipulation by "nonstate actors."

Given the assumed need to fight in the Baghdads of the future, the question for the military quickly becomes a practical one: How to deal with cities as battle zones? That's where DARPA and other DoD dreamers come in. According to DARPA's 2004 report, what was needed were "new systems and technologies for prosecution of urban warfare . . . [and] new operational methods for our sol-diers, Marines, and special operations forces."

These days, DARPA and other Pentagon ventures like the Small Business Innovation Research Program (through which the "DoD funds early-stage R&D projects at small technology companies") and the Small Business Technology Transfer Program (where funding goes to "cooperative R&D projects involving a small business and a research institution") are awash in "urban operations-oriented programs." These go by the acronym of UO and are designed to support tomorrow's interventions and occupations. The director of DARPA's Information Exploitation Office put it this way: "[They are aimed at] conflicts in high density urban areas . . . against enemies having social and cultural traditions that may be counter-intuitive to us, and whose actions often appear to be irrational because we don't understand their context." Such programs include a wide range of efforts to visualize, map out, and spy on the global megafavelas that the United States had, until now, largely scorned and neglected. A host of unmanned vehicles are also being readied for surveillance and combat in these future "hot zones," while all sorts of lethal enhancements are in various stages of development to enable American troops to more effectively kick down the doors of the poor in 2025.

SPIDER-MEN AND EXPLODING FRISBEES

So let's take a closer look at the Pentagon's current UO-oriented systems under development:

> *VisiBuilding.* This program is aimed at addressing "a pressing need in urban warfare: seeing inside buildings"—by developing technology that will allow U.S. forces to "determine building layouts, find anomalous quantities of materials," and "locate people within the building." Think of it as a high-tech military Peeping Tom system that lets U.S. troops spy inside foreign homes and make judgments about whatever they might deem "anomalous" inside. While VisiBuilding is being perfected, troops will have to be content with "Radar Scope," which allows them to "sense through 12 inches of concrete to determine if someone is inside a building."

Camouflaged Long Endurance Nano Sensors. This "real-time ultra-wideband radar network . . . will detect, classify, localize, and track dismounted combatants . . . in urban environments." In translation, a system of palm-sized, networked sensors will monitor an area, day in, day out, for weeks at a time. This is what DARPA likes to call "persistent surveillance." The U.S. military has headed down this particular surveillance path before via the ill-fated McNamara Line of the Vietnam era, a collection of devices that proved incapable of differentiating between armed combatants and civilians. There's little reason to believe anything will change in future urban slums, despite the increasing technological sophistication of such systems.

UrbanScape. This program aims "to make the foreign city as 'familiar as the soldier's backyard'" by providing "the warfighters patrolling an urban environment with an up-to-date, high resolution model of the urban terrain that can be viewed, manipulated and analyzed." Specially outfitted unmanned aerial vehicles (UAVs) and Humvees will gather data about a target city and then translate it into 3D visuals. These images will then be available to troops for use in navigating through and conducting combat operations in tomorrow's labyrinthine slums.

Urban Photonic Sandtable Display. This program "seeks to develop a large holographic display to facilitate rapid and clear communication of intelligence for team-based mission planning and rehearsal, visualization and interpretation of real-time data, and training." In other words, a 3D rendering of a potentially hostile urban environment—a high-resolution holographic image with extremely realistic detail—is to be created, allowing an entire team of soldiers to view it at the same time.

Heterogeneous Urban RSTA Team. With the apt acronym HURT, this program will network together a squadron of small, low-altitude UAVs sending video footage to handheld devices for the immediate use of urban RSTA (reconnaissance, surveillance, and target acquisition) troops. This high-tech system is designed, according to DARPA's director, Anthony J. Tether, to provide U.S. forces with "unprecedented awareness that enables them to shape and control [a] conflict as it unfolds." It is meant to improve the odds when American counterinsurgency warriors take on "warfighters in a MOUT [Military Operations on Urban Terrain] environment" or any ragtag slum militia of tomorrow.

The air force is, in turn, seeking the "ability to continuously track, tag, and locate (TTL) asymmetric threats in urban environments using sensors across the tiers of airborne assets." In other words, a slew of UAVs loitering long-term above hostile cities and slums, ready at a moment's notice to spot a target and begin tracking it. Such "targets" might be "commercial vehicles" or individuals identified through a "hyperspectral imaging HSI video camera" that allows for "the frequency spectrum of clothes, hair, and skin [to] be exploited," thus providing "targeting level accuracy to weapon delivery assets." Think of it as *the* high-tech urban hunter-killer system for the neocolonial future. The air force not only sees this as a way to target and kill "anti-occupation forces" in Baghdad in 2025 but also envisions it doing double duty in the homeland, where, they say, "law enforcement require[s] urban target tracking."

Nano Air Vehicle. Imagine a world in which mechanical gnats infest a city, buzzing through people's homes, filming whatever they choose with tiny cameras, and transmitting the data back to U.S. troops. This program aims to "develop and demonstrate an extremely small (less than 7.5 cm), ultra-lightweight (less than 10 grams) air vehicle system . . . to provide the warfighter with unprecedented capability for urban mission operations."

Multi Dimensional Mobility Robot (MDMR). In a nutshell, this robot "will traverse complex urban terrain."

Micro Air Vehicle (MAV). A small, vertical takeoff and landing UAV that will be "employable in a variety of warfighting environments" including "urban areas."

Urban Hopping Robots. An intriguing but shadowy program whose project manager, Michael Obal, declined to answer inquiries about the project. In an e-mail, Jan R. Walker of DARPA's External Relations office indicated that there is "very limited information available on the Urban Hopping Robots program" but suggested that the "program is developing a semi-autonomous hybrid hopping/articulated wheeled robotic platform that could adapt to the urban environment in real-time and provide the delivery of small payloads to any point of the urban jungle while remaining lightweight, small to minimize the burden on the soldier." The proposed hopping robot, she noted, "would be truly multi-functional in that it will negotiate all aspects of the urban battlefield to deliver payloads to non-line-of-sight areas with precision."

Z-Man. Trademark infringement was probably the only thing that stopped this DARPA program from being called the "Spider-Man Project." Basically, Z-Man seeks to "develop climbing aids that will enable an individual soldier to scale vertical walls constructed of typical building materials without the need for ropes or ladders." The Pentagon is aiming to find methods similar to those employed by "geckos, spiders, and small animals [to] scale vertical surfaces, that is, by using unique biological material systems that enable controllable adhesion." This weaponized wall crawler, assumedly capable of creeping into some 2025 apartment window in Baghdad, Beirut, or Karachi "carrying a combat load," is definitely not meant to be your friendly neighborhood Spider-Man.

Modular Disc-Wing (Frisbee) Urban Cruise Munition. Yes, you read it right; the air force has green-lighted Triton Systems (which received over $9.5 million from the DoD in 2006) to create "a MEFP [Multiple Explosively Formed Penetrator]-armed Lethal Frisbee UAV." That is, a flying disk that will "locate defiladed combatants in complex urban terrain" and annihilate them using a bunker-buster warhead. Unlike your run-of-the mill Wham-O, however, this "frisbee" will probably be thrown using a device resembling a skeet launcher.

Close Combat Lethal Recon. This deadly, loitering explosive expressly for use in urban landscapes is known in DARPA's budgetary documents as the "Confirmatory Hunter-Killer System." It will expand a soldier's killing zone by reaching "over and around buildings, onto rooftops, and into open building portals." Think of it as a *smart* grenade or, according to DARPA director Tether, "a tube-launched cruise munition that can be used by a dismounted infantryman in an urban area to attack a target, perhaps spotted by a UAV, which is beyond his line of sight. It's like a small mortar round with a grenade-size explosive in it. A fiber-optic line unreels from its back end and provides the data link that allows the soldier to see the video from the munition's camera and to fly it into the target."

TRAINING FOR TOMORROW'S URBAN OCCUPATIONS

A glance at Pentagon expenditures makes clear the heavy emphasis on training the men and women who are slated to use DARPA's high-tech urban weapons against slum dwellers in the coming years. In March 2006, the army signed a nearly $25-million contract "for construction of a combined arms collective training facility/urban assault complex" at Fort Carson, Colorado. In August, the navy inked an $18.5-million deal for the "design and construction of a combined arms military operations in urban terrain facility" at Twentynine Palms, California. In September, the army approved a contract for the construction of the Urban Assault Course at Fort Jackson, South Carolina. In November, the navy awarded a $12,500,000 contract for construction of a "Special Operations Force Military Operations on Urban Terrain Training Complex" at San Clemente Island, California. And in December 2006, the army agreed to pay $11,838,998 for a new "Military Operations Urban Terrain Facility" for Fort Irwin, California. In 2007, such efforts were ramped up further. In April, the navy, on behalf of the marines, continued its efforts at Twentynine Palms, California, signing a startling $461,635,454 contract for the creation of a "Combined Arms Military Operations in Urban Terrain (CAMOUT) training system" to "provide a realistic environment to support a variety of training tasks related to the deployment and maneuvering in an urban seeting [*sic*] of a Marine Expeditionary Battallion [*sic*] and its constituent elements."

The Pentagon has even exported its urban warfare training centers to sites closer to tomorrow's prospective targets, such as the army's custom-made MOUT facilities at Bagram Air Base, Afghanistan, and at Camp Buehring, Kuwait. In November 2006, the army also awarded General Dynamics a $17-million contract to construct an urban combat training site as part of the King Abdullah II Special Operations Training Center in Jordan—a facility that will, according to an army spokesman, be available to "all friendly nations that support the War on Terror."

AMERICAN TERMINATORS VS. DRUG-DEALING, SERIAL-KILLER GUERRILLAS

As both the high-tech programs and the proliferating training facilities suggest, the foreign slum city is slated to become *the* bloody battlespace of the future. Curiously, the Pentagon's conceptualization of urban space mimics Hollywood's *Escape from New York*-meets-*Bladerunner*-meets-*Zulu*-meets-*Robocop*-style vision of the third-world city to come.

For example, the U.S. Navy/Marine Corps launched a program seeking to develop algorithms to predict the criminality of a given building or neighborhood. The project, titled Finding Repetitive Crime Supporting Structures, defines cities as nothing more than a collection of "urban clutter [that] affords considerable concealment for the actors that we must capture." The "hostile behavior bad actors," as the program terms them, are defined not just as "terrorists," today's favorite catch-all bogeymen, but as a panoply of nightmare archetypes: "insurgents, serial killers, drug dealers, etc." For its part, the army's recently revised Urban Operations manual offers an even more extensive list of "persistent and evolving urban threats," including regional conventional military forces, paramilitary forces, guerrillas, and insurgents as well as terrorists, criminal groups, and angry crowds. Even the possible threat posed by computer "hackers" is mentioned.

To deal with such dystopian megacities—where serial killers, druglords, hackers, and urban guerrillas may have joined forces— only the Terminator will do. And indeed, DARPA is intent on developing a program worthy of a direct-to-video sci-fi thriller. In a recent solicitation, it offered a vision of a human-robot military SWAT team busting down doors in a favela of the future:

> The challenge is to create a system demonstrating the use of mul-tiple robots with one or more humans on a highly constrained tactical maneuver . . . One example of such a maneuver is the through-the-door procedure often used by police and soldiers to enter an urban dwelling . . . [where] one kicks in the door then pulls back so another can enter low and move left, followed by another who enters high and moves right, etc. In this project the teams will consist of robot platforms working with one or more human teammates as a cohesive unit. The robots should be under autonomous control rather than remote/teleoperated.

This urban scenario of tomorrow already seems well launched. The military has, in fact, been obsessed with the idea of sending to war heavily armed, tele-operated robots—such as the Special Weapons Observation Reconnaissance Detection System, or SWORDS Talon, a small, all-terrain tracked vehicle, used by the U.S. military since 2000, that can be outfitted with M240 or M249 machine guns, Barrett 50-caliber rifles, 40 mm grenade launchers, and antitank rocket launchers. Noah Shachtman, writing for *Wired*'s Danger Room blog, reported in 2007 that, for the first time, SWORDS robots armed with M249 machine guns had been deployed to Iraq. They had yet to fire their weapons by August, "but," said the SWORDS program manager Michael Zecca, "that'll be happening soon."

PENTAGON TO GLOBAL CITIES: DROP DEAD

In late 2006, the Pentagon's U.S. Joint Forces Command engaged in a $25-million, thirty-five-day, computer-based simulation exercise involving more than 1,400 soldiers, marines, airmen, and sailors. A year in the making, "Urban Resolve 2015" had one simple goal—to test concepts for future "combat in cities"—and, not surprisingly, it was set in Baghdad in 2015. An article put out by the Pentagon's American Forces Press Service was quick to point out, however, that this virtual exercise could really be taking place in "any urban environment." And the reason why was clear in the words of Dave Ozolek, the executive director of the Joint Futures Lab at the Joint Forces Command. Urban zones, he said, are "where the fight is, that's where the enemy is, that['s] where the center of gravity for the whole operation is."

While the Joint Forces Command may already be war-gaming the 2015 Battle for Baghdad, right now it looks like the U.S. military will have trouble hanging on there for even a few more years. Still, if present plans become reality, odds are U.S. military planners will be ready once again to occupy some city, in some fashion, come 2015 or 2025. In the future, as the army's new urban operations manual puts it, "every Soldier—regardless of branch or military occupational specialty—must be committed and prepared to close with and kill or capture threat forces in an urban environment."

The way the Pentagon seems to envision the future, its human-robot expeditionary forces will spend increasing amounts of time dropping in on third-world superslums armed not only with heavy weaponry but also with gadgets galore. They will be able to read instant 3D maps of the buildings they're approaching and watch real-time video of the most intimate activities in the urban zone they've been tasked to subdue.

As tiny flying UAVs blanket an impoverished neighborhood, a squad of Special Ops Spider-Men and Gecko warriors will crawl and slither up apartment-building walls, while teams of robots simultaneously hop through first-floor windows and Terminator-Human teams kick down front doors to capture an enemy drug kingpin. Nearby "angry crowds" of politically minded youth will be engaged by heavily armed tele-operated SWORDS Talon robots, while a few up-armored cyborg troops, at a safe distance, will fire their loitering smart grenades at a gathering crowd of armed slum dwellers who believe themselves well hidden and protected in nearby alleyways.

Of course, no matter the fantasies of Pentagon scientists and planners, such futuristic solutions will not replace U.S. reliance on massive firepower, even in labyrinthine cities, as was true with Tokyo during World War II, Pyongyang during the Korean War, Ben Tre in Vietnam, and Fallujah during the current war in Iraq. As Major Tim Karcher, the operations officer for the army's Task Force 2-7 Cavalry, recalled of the American assault on Fallujah in November 2004, "We sat there for a good six or seven hours . . . watching . . . this death and destruction rain down on the city, from AC-130 [gunship]s to any kind of fast-moving aircraft, 155 [millimeter] howitzers. You name it, everybody was getting in the mix."

Given the military's fear of sending large numbers of American troops into the enemy-friendly landscape of the urban megaslum, where significant casualties are almost unavoidable, this form of Pentagon-preferred urban renewal is unlikely to be replaced, no matter what technologies come down the pike.

THE MILITARY AND THE METROPOLIS

In 2005, DARPA's deputy director, Robert F. Leheny, announced:

> Over the past year, we've begun more than 35 study efforts on ideas received in response to our Agency-wide Urban Operations broad agency announcement. And we've launched a number of new programs across all DARPA offices and programs. We're targeted to spend more than $340 million on urban operations programs in the coming year and we anticipate these programs will grow to over $400 million in the out years.

Cities are obviously on the Pentagon's hit list. Today, it's Baghdad; tomorrow, it could be Accra, Bogotá, Dhaka, Karachi, Kinshasa, Lagos, Mogadishu (again), or even a perennial favorite, Port-au-Prince. Yet even with high-tech exploding frisbees, spiderman suits, and numerous urban training facilities coming online, the outlook for the U.S. military is not upbeat.

In the wars begun since the U.S. high command moved into its own self-described virtual "city"—the Pentagon—a distinct inability to decisively defeat any but its weakest foes has been in evidence. Korea in the early 1950s, Vietnam in the 1960s and 1970s, Lebanon in the early 1980s, and Somalia in the early 1990s were all failures. More recently, victory in Afghanistan has proved worse than elusive and a ragtag insurgency in Iraq has fought the Pentagon's much-hyped technological dominance and superior firepower to a standstill. While able to cause massive casualties and tremendous destruction, the Pentagon war machine has proven remarkably ineffectual when it comes to achieving actual victory.

Now, the Pentagon has decided to prepare for a fight with a restless, oppressed population of slum dwellers already 1 billion strong and growing at an estimated rate of 25 million people per year. To take on even lone outposts in this multitude—like any of the 400 cities of over 1 million people that exist today or the 150 more estimated to be in existence by 2015—is a recipe for both carnage and quagmire.

PART VIII

COMPLEX
CONCLUSIONS

Defending the American Homeland 1993–2003," a document from the U.S. Air Force's Counterproliferation Center, notes that "national security" was once the concern of "the Departments of Defense, State, and the Intelligence Community." By 2003, however, "national security" had been joined by "homeland security" and picked up a few more concerned agencies and departments of various sorts. To be exact, according to the report, there were "57 federal agencies, 50 states, 8 territories, and 3,066 counties involved . . . 87,000 different governmental jurisdictions that will play roles in homeland security." Beyond these, the document explained, "the private sector will play a major role in homeland security, and this role will not be limited to logistical support. Corporate America will be required to play a major operational role in homeland security."

The corporations that have become involved in locking down the "homeland" are often, not surprisingly, the same ones that have long fueled the military-corporate complex. For instance, in 2005, it was announced that Lockheed had been awarded a $212-million contract to build a one-thousand-camera, three-thousand-electronic-sensor surveillance and security system for New York City's subways, commuter railroads, bridges, and tunnels. By late 2006, the price tag for that same system had already risen to $280 million.

Technologies used in war zones like occupied Iraq have also found their way back to the homeland. In 2006, for instance, it was revealed that the Los Angeles County Sheriff's Department had begun testing the use of remote-controlled surveillance drones. Later that year, according to the Associated Press, air force secretary Michael Wynne decided to invert the normal weapons testing order, professing his belief that futuristic pain-producing weapons "such as high-power microwave devices should be used on American citizens in crowd-control situations before they are used on the battlefield." This gives new meaning to the phrase "bringing the war home."

25

THE HOMELAND SECURITY STATE

In the wake of September 11, 2001, the United States was almost instantly transformed from a nation into a "homeland" and a previously diminutive arm of the Complex—the domestic security component—began to grow at an exponential rate. Since then, the military has increasingly come to see the United States as akin to one of its garrisoned, if not fully occupied, nations overseas. But like so much else in the Complex, the creation of a "homeland security state" isn't strictly a military effort. As historian Andrew Bacevich, a West Point graduate who spent twenty years in the army, noted:

> The question that arises is whether, in fact, we're not already experiencing what is in essence a creeping coup d'état. But it's not people in uniform who are seizing power. It's militarized civilians, who conceive of the world as such a dangerous place that military power has to predominate, that constitutional constraints on the military need to be loosened. The ideology of national security has become ever more woven.

These militarized civilians—in the government, among the public, and in the corporate sector—are busy expanding and further empowering the Complex within the United States in new and profound ways.

The new homeland-security complex they are creating exhibits many of the characteristics of its big brother. Its civilian center-piece, the Department of Homeland Security (DHS), has its own DARPA-like blue-skies outfit: the Homeland Security Advanced Research Projects Agency, or HSARPA. The DHS also now has ties to Hollywood—as exemplified by its behind-the-scenes assistance to the CBS television shows *CSI: Miami* and *NCIS* as well as the 2004 film *The Terminal,* featuring military-entertainment star Tom Hanks (from such Pentagon-aided movies as *Apollo 13, Saving Private Ryan,* and *Forrest Gump*). In 2006, the DHS's Border Patrol began sponsoring a NASCAR team to help its recruiting efforts. It even sent *troops* (teams of Border Patrol agents), just like big brother, to aid in the occupation of Iraq.

Of course, the new homeland-security complex has also created its own lucrative revolving door. In 2006, Eric Lipton of the *New York Times* reported that at least "90 officials at the Department of Homeland Security or the White House Office of Homeland Security—including the department's former secretary, Tom Ridge . . . are executives, consultants or lobbyists for companies that collectively do billions of dollars' worth of domestic security business."

By 2006, the "homeland security" field had become not only a major component of the Complex but a mega-industry in its own right (generating $59 billion that year). It instantly dwarfed such long-established and lucrative enterprises as the motion-picture and music industries ($40 billion each per year). The biggest customer, not surprisingly, is the Department of Homeland Security, created in 2002 as an amalgam of twenty-two agencies and bureaus (and portions thereof), ranging from the Immigration and Naturalization Service and the Animal and Plant Health Inspection Service to the Federal Emergency Management Agency (FEMA), the Secret Service, and the Coast Guard. It now has 180,000 employees, working in four major directorates (Border and Transportation Security, Emergency Preparedness and Response, Science and Technology, and Information Analysis and Infrastructure Protection).

With a self-professed mission to "secure America" and "protect against and respond to threats and hazards to the nation"—a job that an actual Department of *Defense* might otherwise be tasked to do—the DHS acts as a second Pentagon within the Complex. According to data compiled by Veronique de Rugy of the American Enterprise Institute, "homeland security" spending by the U.S. government increased exponentially from $9 billion in 1995 to $58.2 billion in 2007. (That includes a 246 percent increase since 2001.) In 2002, the year it was established, the DHS was allocated $14.1 billion. In 2005, it was receiving double that—$28.9 billion. By 2007, its share of appropriations reached $42.7 billion. For 2008, President Bush requested more than $46 billion, a 7 percent increase.

Such growth is likely to continue, according to Brian Ruttenbur of the investment firm Morgan Keegan. In 2006, he told the *Christian Science Monitor* that he expected "the DHS's budget to increase by 5 to 7 percent annually over the next 10 years." Industry-watcher Homeland Security Research of Washington, D.C., expected "homeland security spending to nearly double by 2010." And the Civitas Group, "a strategic advisory and investment firm serving the homeland and national security markets," predicted in 2007 that the U.S. security market would be worth $140 billion over the next five years.

With this kind of money up for grabs, it's hardly surprising that the homeland-security complex is composed, in large part, of familiar faces from the Complex. "We look at the changing budgets for defense- and space-related activities," Gordon McElroy, the vice president of Lockheed Martin's Intelligence and Homeland Security Systems division, noted in 2005 and see that the "growth engine is going to be in homeland security and law enforcement." The company's Web site gave an indication of exactly why "Lockheed Martin is a partner with our government in the Homeland Security," namely the "opportunities in state and local governments as well as the private sector, including the 17 key sectors that support national security. These range from transportation to finance and from agriculture to the chemical industry." Such "opportunities" have been a boon to big contractors. According to

Dan Verton, a former Marine Corps intelligence officer and founder of the Web site Homeland Defense Week, "The big winners of the homeland security windfall have, as usual, been the big contractors." The accompanying chart lists a selection of contractors, their rankings, and total contract dollars from the DHS and DoD, respectively, in 2006.

Company	DHS 2006 ranking	DHS dollars	DOD 2006 Ranking	DoD dollars
Fluor Corporation	1	1,504,817,784	n/a	278,165,675
Shaw Environmental	2	852,205,338	62	519,041,459
Bechtel	3	471,243,361	27	1,264,475,040
CH2M Hill Constructors	4	436,537,706	86	347,963,421
L-3 Communications	9	270,639,463	7	5,197,490,394
Science Applications International	14	149,190,149	10	3,210,604,531
Northrop Grumman	19	110,937,357	3	16,627,067,499
Booz Allen Hamilton	20	108,812,970	28	1,245,215,183
Lockheed Martin	28	74,620,135	1	26,619,693,002
General Dynamics	35	56,883,056	4	10,526,161,839
Dell Computer	45	49,837,488	50	636,343,593
The Mitre Corporation	58	38,479,032	48	652,276,956
Boeing	62	36,480,266	2	20,293,350,668

Defense giant Raytheon ($10.1 billion in DoD dollars in 2006), for example, "did $38 million in homeland security business in 2001 [and] $106 million in 2002." By 2005, its homeland security sales had reached "about $445 million." Similarly, a Northrop Grumman advertisement boasted that the company's "homeland security revenue" had "soared from $500 million in 2003, to approximately $1 billion in 2005." In 2006, both companies went head-to-head with fellow military-corporate stalwarts, including Boeing and Lockheed, for a DHS border security contract worth about $2.5 billion over four years—which Boeing eventually won.

With the big-name military contractors have come familiar-sounding scandals. For instance, according to the *Washington Post*, Boeing leaped into the homeland security market by

> installing explosive-detection systems at more than 400 airports in less than six months following the Sept. 11, 2001, terrorist attacks. But that contract was criticized by the Homeland Security's inspector general's office, which found that Boeing received $49 million in excess profit on a deal that was supposed to be worth $508 million but ballooned to $1.2 billion. Investigators also found that Boeing had subcontracted 92 percent of the work, and that the machines had high false-alarm rates.

In 2006, according to a bipartisan congressional report, thirty-two separate DHS contracts, worth a total of $34 billion, "experienced significant overcharges, wasteful spending, or mismanagement." The report went on to note that from 2003 to 2005, the total value of contracts awarded without full competition reached $5.5 billion—a 739 percent increase over a mere three years.

Familiar stories of profligate, pork-laden spending also surfaced. The *Washington Post* reported that a review by the Defense Contract Audit Agency of a DHS contract with Pentagon supplier NCS Pearson to hire airport screeners "uncovered at least $297 million of questionable costs, including luxury hotel rooms." Meanwhile, it was also reported that two "TSA [Transportation Security Administration] employees used government purchase cards to buy $136,000 worth of personal items, including leather briefcases."

NORTHCOM AND THE ALPHABET SOUP OF AGENCIES

Military contractors aren't the only mainstays of the homeland-security complex. The U.S. military, itself, is playing a major role. In 2002, the Pentagon established the U.S. Northern Command (NORTHCOM), whose area of operations is "America's homefront." NORTHCOM is much more forthright about what it supposedly doesn't do than what it actually does. Its Web site repeatedly notes that NORTHCOM is not a police auxiliary and that the Reconstruction-era Posse Comitatus Act prevents the military from meddling much in domestic affairs. Nevertheless, NORTH-COM readily, if somewhat vaguely, admits to being involved in a "cooperative relationship with federal agencies" and "information-sharing" among organizations. NORTHCOM's initial commander, General Ralph "Ed" Eberhart, who, the *Wall Street Journal* noted, was the "first general since the Civil War with operational author-ity exclusively over military forces within the U.S," was even more blunt when he told PBS's *Newshour,* "We are not going to be out there spying on people. We get information from people who do."

Those "people who do" are increasingly moving into NORTH-COM's neighborhood. NORTHCOM, it turns out, is "co-located" with the North American Aerospace Defense Command (better known as NORAD) at Peterson Air Force Base in Colorado Springs, Colorado, only a short drive away from Denver. In 2005, the *Wash-ington Post* reporter Dana Priest broke the story of the CIA's plans "to relocate the headquarters of its domestic division, which is responsible for operations and recruitment in the United States, from the CIA's Langley headquarters to Denver." The next year, the *Post*'s William Arkin announced that the National Security Agency was "in the process of building a new warning hub and data warehouse in the Denver area, realigning much of its work-force from Ft. Meade, Maryland to Colorado."

Other agencies have also sent agents to Colorado to work with NORTHCOM. In a 2005 speech, NORTHCOM's second com-mander, Admiral Timothy J. Keating, shed light on the agencies that were collaborating with his command.

At our headquarters in Colorado Springs, we have a Combined Intelligence and Fusion Center—a unique combination of tal-ented professionals and sophisticated capabilities focused on sharing information and analysis with intelligence and law enforcement agencies . . . the whole gamut: FBI to CIA; NSA to the Coast Guard [Intelligence]; from the National Counterterrorism Center, newly stood up in Virginia, to the [Pentagon's C]ounterintelligence [F]ield [A]ctivity that includes both American and Canadian experts.

It's a pretty good bet that NORTHCOM maintains relationships with the other twelve members of the U.S. Intelligence Community (IC), including: Air Force Intelligence, Army Intelligence, the Defense Intelligence Agency, the Department of Energy, the Department of Homeland Security, the Department of State, the Department of the Treasury, the Drug Enforcement Administration, Marine Corps Intelligence, the National Geospatial-Intelligence Agency, the National Reconnaissance Office, and Navy Intelligence. These agencies, in turn, maintain relations with executive and legislative branch intelligence entities, including the National Security Council, the President's Foreign Intelligence Advisory Board, President's Intelligence Oversight Board, the Office of Management and Budget, the Senate Select Committee on Intelligence, and the House Permanent Select Committee on Intelligence.

This already massive list of Intelligence Community organizations isn't even comprehensive. There's a seemingly endless supply of half-known, and little understood, if well-acronymed, intelligence/military/security-related offices, agencies, advisery organizations, and committees such as the Counterintelligence Field Activity, or CIFA (mentioned by Keating), the Defense Airborne Reconnaissance Office (DARO), the Department of Defense's own domestic police corps: the Pentagon Force Protection Agency (PFPA), and the Intelligence's Community's internal watchdog, the Defense Security Service (DSS). Additionally, the IC has its own shadowy DARPA-like agency, originally known as Advanced Research Development Activity (ARDA), now known as the Disruptive Technology Office.

Another known member of this intelligence alphabet soup is the Office of the National Counterintelligence Executive (more familiarly known as NCIX)—an organization whose main goal is "to improve the performance of the counterintelligence (CI) community in identifying, assessing, prioritizing and countering intelligence threats to the United States." To help accomplish this task, NCIX, in addition to providing "analysis for the White House, Congress, the Department of Defense, the Intelligence Community, and many other government departments and agencies," offers downloadable "counterintelligence and security awareness posters." One poster features the text of the First Amendment to the Constitution (". . . Congress shall make no law . . . prohibiting the free exercise thereof; or abridging the freedom of speech . . .") and the likeness of Thomas Jefferson, but with an addendum that reads: "American freedom includes a responsibility to protect U.S. security—leaking sensitive information erodes this freedom."

A NCIX poster, released in 2006, announces: "Report suspicious behavior, and you, too can become a CI Super Hero!" Still another NCIX poster looks as if it had arrived through a time warp straight from the old East Germany: "America's Security is Your Responsibility. Observe and Report." While NCIX is an obscure agency, its decision to improve on the First Amendment and a fundamental American freedom appears indicative of where our homeland security state is heading.

OTHER HOMELAND HIJINX

In January 2005, the *New York Times* reported that as part of the "extraordinary army of 13,000 troops, police officers and federal agents marshaled to secure the [presidential] inauguration," the Pentagon had deployed "super-secret commandos . . . with state-of-the-art weaponry" in the nation's capital. This was done using government directives that undercut the Posse Comitatus Act of 1878. According to the *Times,* the black-ops cadre, based at the ultrasecretive Joint Special Operations Command at Fort Bragg, North Carolina, were operating under "a secret counterterrorism program code-named Power Geyser." The program was brought to light in *Code Names,* a book by the former army intelligence analyst William Arkin, who revealed that "special-mission units [are being used] in extra-legal missions . . . in the United States" on the authority of the Department of Defense's Joint Staff and with the support of the DoD's Special Operations Command and NORTHCOM.

Power Geyser isn't an anomaly. In fact, since 1989, in the name of the War on Drugs, there has been a multiservice command (made up of approximately 160 soldiers, sailors, marines, airmen, and Department of Defense operatives) known as Joint Task Force Six (JTF-6), providing "support to federal, regional, state and local law enforcement agencies throughout the continental United States." Just how many such commando squads exist and exactly what they are doing can't be known for sure since spokespersons for the relevant military commands refuse to offer comment and Pentagon spokesman Bryan Whitman would only say that at "any given time, there are a number of classified programs across the government" and that Power Geyser "may or may not exist."

The military has been creeping into civilian life in the United States in all sorts of ways. In 2006, Arkin reported that a secret initiative begun in 1999, known as "Able Danger," used "state-of-the-art information technology and advanced analytic techniques . . . to try [to] 'map out' the al-Qaeda network." Not only was the effort, which eventually took up residency at a Raytheon facility, a failure, but, as Arkin noted, one part of the project "illegally gathered material on American citizens" that found its way into U.S. government files.

In 2003, Torch Concepts, an army subcontractor, was given Jet-Blue's entire 5.1-million-passenger database, without the knowledge or consent of those on the list, for data mining—a blatant breach of civilian privacy that the army nonetheless judged *not* to violate the federal Privacy Act. That same year, the *Wall Street Journal* reported that the Office of Naval Intelligence had badgered the U.S. Customs Service to hand over its database on maritime trade. At first, the Customs Service resisted the navy's efforts, but eventually it caved in to military pressure. In an ingenuous message sent to the *Wall Street Journal,* the commissioner of the Department of Homeland Security's Bureau of Customs and Border Protection, Robert C. Bonner, excused handing over the civilian database by stating that he had received "navy assurances that the information won't be abused."

In 2004, the FBI, in conjunction with the Department of Homeland Security, launched its October Plan. According to a CBS news report, this program consisted of "aggressive—even obvious—surveillance techniques to be used on . . . people suspected of being terrorist sympathizers, but who have not committed a crime," while other "'persons of interest,' including their family members, m[ight] also be brought in for questioning." In late 2005, the NSA warrantless wiretapping scandal broke, followed, in early 2006, by the disclosure of information about "secret rooms" that were reportedly set up inside AT&T switching centers. The rooms, perhaps fifteen to twenty around the country, were created by the military apparently to allow the National Security Agency to monitor "electronic voice and data communications" moving through the AT&T facilities. In 2006, the *New Scientist* also discovered that the NSA was bankrolling research to collect all the personal information available on MySpace.com and other social-networking sites. In 2007, it came to light that the warrantless surveillance of e-mails and phone calls that was revealed in 2005 was, according to the *Washington Post*, "part of a much broader operation" of secret spying.

The 2005 disclosure, by the ACLU, of documents that showed "the FBI had investigated several activist groups, including People for the Ethical Treatment of Animals (PETA) and Greenpeace, supposedly in an effort to discover possible ecoterror connections," was soon followed by a military analog. In 2006, it was revealed that army analysts from CIFA, the Pentagon's "force protection" unit that tracks threats against military installations had "collected information on nearly four dozen antiwar meetings or protests, including one at a Quaker meetinghouse in Lake Worth, Fl[orida], and a Students Against War demonstration at a military recruiting fair at the University of California, Santa Cruz." They had even spied on peace activists protesting outside the Houston headquarters of Halliburton, the massive defense contractor once headed by Vice President Dick Cheney. According to a *Newsweek* exposé, a Pentagon official admitted that "the number of reports with names of U.S. persons could be in the thousands," and a 2007 inspector general's report found that the Pentagon improperly retained information it had amassed.

In 2006, it was also revealed that KBR—the former Halliburton subsidiary responsible for building prison facilities at Guantánamo Bay and for a series of scandals stemming from work in the Iraq war zone—received a $385-million contract from the Department of Homeland Security (DHS) to build detention centers, according to the *New York Times*, "for an unexpected influx of immigrants" or "new programs that require additional detention space." The long history of detention measures, from the World War I–era proposal to imprison members of the Industrial Workers of the World for the duration of the conflict, to the forcible relocation and imprisonment of Japanese and Japanese-Americans during World War II, to the Vietnam-era plans to round up and jail radicals in the event of a national emergency and conduct mass detentions in the face of possible urban insurrections, no doubt caused former Pentagon analyst Daniel Ellsberg to worry about the implications. "Almost certainly this is preparation for a roundup after the next 9/11 for Mid-Easterners, Muslims and possibly dissenters," he said. "They've already done this on a smaller scale, with the 'special registration' detentions of immigrant men from Muslim countries, and with Guantánamo."

In 2007, NORTHCOM requested Pentagon permission to "establish its own special operations command for domestic missions," wrote Arkin—essentially to create an official homeland commando unit. The trouble, as Arkin observed, was that NORTHCOM was "already doing what it ha[d] requested permission to do." Back in 2002, the year it was created, NORTHCOM actually set up a "Compartmented Planning and Operations Cell (CPOC) responsible for planning and directing a set of 'compartmented' and 'sensitive' operations on U.S., Canadian and Mexican soil." From the very beginning, it seems, NORTHCOM's area of operations was just another war zone in which to deploy troops.

Like its military counterpart, the Department of Homeland

Security has its own armed squads operating in the United States. On July 4, 2007, Visual Intermodal Protection and Response, or VIPR, teams—small groups of U.S. air marshals, bomb detection specialists, and others from DHS's Transportation Security Administration—fanned out across locations in and around Washington, D.C., Boston, Houston, Los Angeles, San Francisco, and other major U.S. cities. During the previous eighteen months, according to the TSA, they had "partnered with local law enforcement" to conduct eighty-four VIPR missions.

Partnerships with other security forces are—as with its military analog, NORTHCOM—a key to how the DHS does business. In the department's operations center, the "FBI, the CIA, the Secret Service, and 33 other federal agencies each has its own workstation. And so do the police departments of New York, Los Angeles, Washington and six other major cities." There are even large signs on walls and doors that remind operatives: "Our Mission: To Share Information"; and, to facilitate this, local police officers sit just "a step or two away from the CIA and FBI operatives who are downloading the latest intelligence coming into those agencies."

MERCENARY FORCES

Privatization schemes launched in the 1990s, and ramped up since 2001, have radically transformed the nature of the U.S. military and the Complex itself. For instance, by mid-2007, in the midst of a major U.S. troop surge, the *Los Angeles Times* reported that there were more "than 180,000 civilians . . . working in Iraq under U.S. contracts"—a force that outnumbered the actual military troop presence by 20,000. Of them, according to a Government Accountability Office report, 48,000 were employed by private military companies. These mercenary forces, organized by companies like Blackwater USA and DynCorp, have also increasingly shown up in the homeland.

In the aftermath of Hurricane Katrina, in 2005, for instance, journalists Jeremy Scahill, who would go on to write *Blackwater:*

The Rise of the World's Most Powerful Mercenary Army, and Daniela Crespo reported that "armed paramilitary mercenaries from the Blackwater private security firm, infamous for their work in Iraq, are openly patrolling the streets of New Orleans." Blackwater USA personnel told them that they were working for the Department of Homeland Security and sleeping in camps set up by the DHS. The *Washington Post* later reported that Blackwater "earned about $42 million through the end of December [2005] on a contract with Federal Protective Service, a unit of the Department of Homeland Security, to provide security to FEMA sites. Most of the 330 contract guards now working in Louisiana are employed by the company."

Similarly, DynCorp, which took home over $1 billion from the Pentagon in 2006, used Hurricane Katrina as an entrée to the homeland. A major security contractor for the U.S. government in both Iraq (where it became embroiled in scandal, in late 2007, when U.S. government auditors could not account for almost $1.2 billion in U.S. State Department funding it was given to train Iraqi police) and Afghanistan (and subsidiary of defense contractor Computer Sciences Corporation—the eleventh-largest DoD contractor in 2006), DynCorp, according to the *Washington Post,* "conducted its first domestic mission after Hurricane Katrina, setting up housing and offices for police in St. Bernard Parish after the storm and sending a couple of dozen of its private security people to the region." A $250-million contract from the DHS's FEMA followed in 2006. In June 2007, the president of DynCorp International's government services division, Robert Rosenkranz, proposed to the House Homeland Security Subcommittee on Management, Investigations and Oversight that his company provide one thousand of its private agents to augment federal forces patrolling the U.S. border. The proposal was well received by the ranking Republican on the subcommittee, Mike Rogers (Alabama), who maintained, wrote Federaltimes.com, "that security contractors such as DynCorp and Blackwater could help provide more manpower."

START SNITCHING

The ultimate form of privatization in the homeland-security complex might be the civilian-military push to create an army of private citizens willing to spy on their families, friends, and neighbors. The programs are already taking shape.

In 2002, the Bush administration's attempt to induce millions of Americans to conduct surveillance on their fellow citizens through the Terrorism Information and Prevention System, or TIPS, went down in flames in Congress. But that wasn't the end of such programs. TIPS was part of the Citizen Corps, an initiative launched in January 2002 by President Bush and "coordinated nationally by the Department of Homeland Security." Another program, operating under the auspices of the Citizen Corps, survives. USAOnWatch was created, says its Web site, "by the National Sheriffs' Association, in conjunction with several well-known federal agencies," and allows members "of local communities and representatives from businesses, government agencies, and a variety of organizations" to form "partnership[s] with their local law enforcement agency," creating neighborhood surveillance programs. These are not, however, geared only toward thwarting crime but also "address terrorism prevention."

In addition to the creation of an FBI "Web site to report suspected terrorism"—tips.fbi.gov—and a link on the DoD's homepage to "report suspicious activity" to the Pentagon Force Protection Agency, the DHS's Coast Guard launched "America's Waterway Watch." This program relies on "normal citizens" to "logically determine what is suspicious" and report such activities to the security forces. Says its Web site, "'You know you have a job to do,'" continuing, "You know what 'normal activity' is and, conversely, what activity is 'not normal.'"

Similarly *rigorous* standards have been adopted for the air force's domestic surveillance program, Eagle Eyes, an antiterrorism initiative that aims to enlist average citizens in the Global War on Terror. The Eagle Eyes' Web site tells viewers: "You and your family are encouraged to learn the categories of suspicious behavior," and

The approved Eagle Eyes logo. *Image courtesy of the U.S. Air Force.*

it exhorts the public to report in to "a network of local, 24-hour phone numbers . . . whenever a suspicious activity is observed." Just what, then, constitutes "suspicious activity"? Among behaviors that merit the air force's attention are the use of still or video cameras, note taking, making annotations on maps, or using binoculars. (Bird-watchers beware!) The air force advises that the eagle-eyed should be on the lookout for "suspicious persons out of place . . . People who don't seem to belong in the workplace, neighborhood, business establishment, or anywhere else." While the air force admits that "this category is hard to define," it offers a classic you-know-it-when-you-see-it definition: "The point is that people know what looks right and what doesn't look right in their neighborhoods, office spaces, commutes [*sic*], etc, and if a person just doesn't seem like he or she belongs . . ."

COMPLEX CONCLUSIONS

The post-9/11 creation of an entire industry and culture of "homeland security" has ushered in an era of military-industrial transformation. With each passing month, the Complex (and its attendant minicomplexes) grows larger and embeds itself ever more deeply in American society, becoming more like a real version of the Matrix with every passing day.

During Dwight Eisenhower's presidency, Congress wasn't exactly stingy with the "permanent armaments industry" that Ike decried in his farewell address. During his eight years in office, defense spending topped $350 billion. In today's dollars, that number would be roughly $3 *trillion*. However, these days, even that massive early Cold War–driven sum would be dwarfed in less than four years according to the sophisticated analysis of national security spending conducted by Robert Higgs, a senior fellow of the Independent Institute.

Higgs took into account not only Department of Defense and Department of Homeland Security spending but also the "defense-related" portions of budgets for the Departments of State, Energy, and Justice as well as the National Aeronautics and Space Administration (NASA), and the costs of the Department of Veterans Affairs, the Treasury Department's military retirement fund, and the "net interest attributable to past debt-financed defense outlays." He then calculated a total single-year national security budget of $934.9 billion and estimated a true expenditure of between $987 billion and $1.028 trillion in 2007. Given what we know about the Complex's penchant for expansion, we're essentially assured of similar budgets for the foreseeable future.

From the 1940s through his years in the White House, Eisenhower repeatedly decried unrestrained defense spending as a pathway to a "garrison state" where the military would hold extraordinary sway over the nation. Today, the United States has begun to resemble what Eisenhower feared. Having garrisoned the globe, the Complex is returning home in new and unnerving ways.

In the coming years, we're apt to see more previously *foreign-deployed* aspects of the Complex—such as the use of mercenaries, surveillance drones, and computer-aided tracking technology—appear in the United States courtesy of the homeland-security complex. Meanwhile, fueled by wars and occupations abroad, the Complex will also continue to swell. As this book has tried to demonstrate, it is already an entity light-years beyond the size, scale, and scope of Dwight Eisenhower's military-industrial complex, and its most powerful aspects don't have the look of a garrison state at all. From sunglasses to video games, golf courses to doughnuts, hot movies to hot cars, much of the way the Complex manifests itself hardly looks "military" at all.

For most Americans, and many other peoples across the planet, the Complex is a powerful engine that helps to drive our world, a vast system of systems, a real Matrix, hidden in plain sight. The Complex thrives on the very obliviousness of the civilian population to its existence in the world it has made so much its own. But if you look closely, it can suddenly come into focus and be seen almost everywhere: on our TVs and in the movies we watch; in the video games we play and the products we buy; in the coffee we drink and the boots we wear; in the stocks we own and the Web sites we visit; and in almost every other facet of our lives.

What's next? Perhaps one of the many taglines for *The Matrix* says it all: "The Fight for the Future Begins."

BIBLIOGRAPHIC ESSAY

To maximize readability, rather than footnote every reference on the page, I have posted complete citations and additional information online. Readers interested in learning more about the subjects touched upon should go to http:www.americanempireproject.com or http://www.nickturse.com. But even these many notes do not provide citations for all the facts contained in this book.

Most of the information on corporations and their contracts with the Pentagon comes from the Department of Defense's own publications, specifically the yearly procurement reports, summaries, and data files that the DoD releases. Chief among these sources are the massive contractor lists—documents that generally span four to six thousand pages in length. These documents are periodically available from the DoD online but are sometimes inaccessible for large stretches of time. Older procurement documents seem to disappear entirely, leading to dead links that inform viewers that the documents are "not found on this server."

As a result, it seemed pointless to provide citations leading to error messages and documents that might rarely, or never again, be available. It is also worth noting that the DoD does not make it easy to verify the information in its procurement documents, and contractors are generally wary, unwilling, or unable to confirm the dollar amounts, the purpose of the contract(s), or even the existence of the contract(s). Still, despite all these drawbacks, DoD procurement data remains an excellent and underutilized source of information on corporate collaborations with the U.S. military.

Other types of uncited procurement-oriented documents are also key to certain chapters. For instance, chapter 2, "The Military-Academic Complex," relies heavily on the DoD's annually published list of the top one hundred "Research, Development, Test, and Evaluation" (RDT&E) contractors.

Many other chapters utilize information available on or through the DoD's official Web site: http://www.defenselink.mil/. While thoroughly partisan, to say the least, it still offers a wealth of useful information on the U.S. military. Much of the information in this book on individual contracts awarded to corporations for specific tasks comes from regularly released contract data available at that Web site.

Another key source of information used in writing this book, especially chapter 4, "Global Landlord," is the DoD's yearly *Base Structure Report*. Published annually since 1999, this inventory of the Pentagon's property portfolio contains a wealth of information, however limited or restricted, on the size and location of bases and other facilities around the world.

Due to its unique format, chapter 11, "Six Billion Movies and No Separation," contains very few endnotes. This chapter relies heavily on the tremendous amount of information in Lawrence H. Suid's *Guts and Glory: The Making of the American Military Image in Film* (Lexington: University Press of Kentucky, 2002). Another very useful source was Frank McAdams's *The American War Film: History and Hollywood* (Westport, CT: Praeger, 2002). Additionally, the Internet Movie Database (http://www.imdb.com) was extremely useful, as was the University of Virginia's IMDb-dependent "Oracle of Bacon at Virginia" (http://oracleofbacon.org/).

This book owes a great debt to the exceptional body of work by numerous writers and reporters who doggedly cover aspects of the military "beat." On the tech side of the Complex, Noah Shachtman is without parallel. Without his work, my life (and the writing of this book) would have been immensely more difficult. His writings can be found at *Wired*'s Danger Room blog (http://blog.wired.com/defense/). Not only was reporter William Arkin's *Code Names: Deciphering US Military Plans, Programs, and Operations in the 9/11 World* (Hanover,

NH: Steerforth Press, 2005) of great help to this project, but his columns over the years have enabled all of us to get at least a glimpse of what is going on behind a few closed doors in the Pentagon and the Intelligence Community. His informative blog, Early Warning, can be found through the *Washington Post*'s Web site (http://blogs .washingtonpost.com/earlywarning/).

Many older texts influenced my thinking greatly and helped inform my work, among them C. Wright Mills's still-prescient *The Power Elite* (New York: Oxford University Press, 1957); Seymour Melman's *Pentagon Capitalism* (New York: McGraw-Hill, 1970) and *The War Economy of the United States: Readings on Military Industry and Economy* (New York: St. Martin's Press, 1971); James A. Donovan's *Militarism, U.S.A.* (New York: Scribner, 1970); and Sidney Lens's *The Military-Industrial Complex* (Philadelphia: Pilgrim Press, 1970).

More recent texts that were key to my understanding of militarism today include Andrew J. Bacevich's *American Empire: The Realities and Consequences of U.S. Diplomacy* (Cambridge, MA: Harvard University Press, 2002); Noam Chomsky's *Hegemony or Survival: America's Quest for Global Dominance* (New York: Metropolitan Books, 2003); Carl Boggs's *Imperial Delusions: American Militarism and Endless War* (New York: Rowman & Littlefield, 2005); and Chalmers Johnson's unequaled *The Sorrows of Empire: Militarism, Secrecy, and the End of the Republic* (New York: Metropolitan Books, 2004).

Works of a more specific nature that greatly aided my writing include Robert Latham, ed., *Bombs and Bandwidth: The Emerging Relationship between IT and Security* (New York: New Press, 2003); James Der Derian, *Virtuous War: Mapping the Military-Industrial-Media-Entertainment Network* (Boulder, CO: Westview Press, 2001); Ed Halter's *From Sun Tzu to Xbox: War and Video Games* (New York: Thunder's Mouth Press, 2006); and Lauren Gonzalez's encyclopedic articles "When Two Tribes Go to War: A History of Video Game Controversy" and "Redefining Games: How Academia Is Reshaping Games of the Future" (both available via http://www.gamespot.com).

ACKNOWLEDGMENTS

Books do not just happen. This one would never have come into being without the assistance of so many people—chief among them my editor and good friend Tom Engelhardt. Every author should be so lucky to have at least one opportunity to work with him. He is, without exaggeration, the best in the business. But Tom did far more than put in endless hours reading and editing various iterations of this manuscript, making suggestions, offering insightful criticisms, polishing rough edges, saving me from stupid mistakes, and helping me think through hard questions. He quite literally made me a writer. I haven't figured out how to adequately express my gratitude, so all I can say is, "Thanks, Tom."

Plenty of others deserve my thanks, as well. Not only is Sara Bershtel the last, best hope of mainstream publishing, but she was the project's biggest booster from day one, ever giving of time, insight, and advice. I couldn't imagine, let alone ask for, a better publisher. At Metropolitan, Riva Hocherman deserves particular gratitude; Kate Levin, Grigory Tovbis, and especially Megan Quirk aided this project in innumerable ways and have my sincere thanks. While my agent, Melissa Flashman, does the work that would otherwise make my head swim. Thanks, Mel.

I owe a debt of gratitude to both Vicki Haire, for her excellent copyediting skills, and Ellis Levine, who made the legal review go smoother than I imagined possible. I would also like to thank all the people who provided me with leads and insights and took the time to answer my questions—especially the handful of military personnel who were truly helpful in providing information—as I tried to unravel portions of the Complex. In addition, my gratitude goes to those who sent in information and tips to me through TomDispatch.com (please keep 'em coming) and many fellow writers and reporters who have analyzed aspects of this topic and aided my work, directly or indirectly.

I must also thank my parents for providing not only support and encouragement over the years but workspace at the beginning of this project—as well as my in-laws, since these chapters were edited under their roof.

Finally, and most important, I want to thank Tam, who has lived with this project—and all that comes with it—for as long as I have. Not only did she offer insightful comments and innumerable excellent suggestions, lend her editorial and proofreading expertise, and help me think through many issues in the text, but her support, in ways beyond measure, made this book a reality. I cannot begin to express my appreciation and love.

I am ever grateful to all those who helped edit, sharpen, and hone this text. All mistakes are, however, either my own or those of the Department of Defense.

INDEX

Entries in *italics* refer to illustrations.